SpringerBriefs in Applied Sciences and Technology

Computational Intelligence

Series editor

Janusz Kacprzyk, Polish Academy of Sciences, Systems Research Institute, Warsaw, Poland

The series "Studies in Computational Intelligence" (SCI) publishes new developments and advances in the various areas of computational intelligence—quickly and with a high quality. The intent is to cover the theory, applications, and design methods of computational intelligence, as embedded in the fields of engineering, computer science, physics and life sciences, as well as the methodologies behind them. The series contains monographs, lecture notes and edited volumes in computational intelligence spanning the areas of neural networks, connectionist systems, genetic algorithms, evolutionary computation, artificial intelligence, cellular automata, self-organizing systems, soft computing, fuzzy systems, and hybrid intelligent systems. Of particular value to both the contributors and the readership are the short publication timeframe and the world-wide distribution, which enable both wide and rapid dissemination of research output.

More information about this series at http://www.springer.com/series/10618

Ranjit Biswas

Continuous Fuzzy Evaluation Methods: A Novel Tool for the Analysis and Decision Making in Football (or Soccer) Matches

A New Innovative Proposal to FIFA & UEFA

 Springer

Ranjit Biswas
Department of Computer Science
  and Engineering, Faculty of Engineering
  and Technology
Jamia Hamdard University
New Delhi, Delhi
India

ISSN 2191-530X   ISSN 2191-5318   (electronic)
SpringerBriefs in Applied Sciences and Technology
ISSN 2520-8551   ISSN 2520-856X   (electronic)
SpringerBriefs in Computational Intelligence
ISBN 978-3-319-70750-1   ISBN 978-3-319-70751-8   (eBook)
https://doi.org/10.1007/978-3-319-70751-8

Library of Congress Control Number: 2017957839

Printed on acid-free paper

This Springer imprint is published by Springer Nature
The registered company is Springer International Publishing AG
The registered company address is: Gewerbestrasse 11, 6330 Cham, Switzerland

Football Fans of the World

(A New Innovative Proposal to FIFA & UEFA)

Acknowledgements

The author is thankful to the 'Editor in Chief' Prof. Janusz Kacprzyk for his valuable suggestions which have helped to improve the documentation of this book.

Contents

About the Author

Prof. Ranjit Biswas did his B.Sc.(Hons) in Mathematics from Calcutta University, M.Sc. in Mathematics with Computer Science from IIT Indian Institute of Technology (Kharagpur), M.Tech. in Computer Science from IIT (Kharagpur) and Ph.D. in Computer Engineering from Jadavpur University, Calcutta; being holder of top Ranks/AIR all through including the IIT entrance examinations. He has taught more than 36 years in top institutions like Calcutta University, IGNOU, IIT (Kharagpur), Philadelphia University, NIT, etc. He has guided a large number of Ph.D. scholars, M.Tech. Projects/Dissertations in IIT and in various universities in India and abroad in the field of Computer Science. He served as an examiner of many Ph.D. theses of IIT and various top universities in India and abroad in the field of Computer Science. He has published more than 120 research papers, all being in journals of international repute of USA, Germany, France, UK, Bulgaria, Poland, Austria, Czechoslovakia, etc. where most of his published papers are single authored. His main areas of research are Fuzzy Logic, Soft Computing, Big Data, Atrain Distributed System, NR-Statistics, Theory of CIFS, Region Mathematics, etc. He is Member of Editorial Board of 14 foreign journals of high esteem international repute in the area of Computer Science and Engineering. At present, he is Professor and Head of Department of Computer Science and Engineering, Faculty of Engineering and Technology, Jamia Hamdard University, New Delhi-62, India.

Abbreviations and Their Significance

BGS   By a 'Bad Goal-Shot' (BGS) we mean that the goal shot was an attempt for goal, but it is neither a goal nor an MGS. The ball must pass completely over the goal line. It is assumed that the goal shot is valid as per Laws of the Game, even not invalidated due to handball, offside, etc. **Any goal shot, if qualified to be an SCB (as per definition of SCB given in Sect. 7.2), will never be qualified to be a BGS even if it enters into or does not enter into the net of Goal-Area Post**

Board-1   It is the 'Electronic Display Board' in which details of data/information about the latest updated CS values of positive and negative parameters, as well as the latest CS Score of both the teams, are displayed during 90 minutes of play

Board-2   It is used during CPS play. It is the 'Electronic Display Board' in which details of data/information about the latest updated CS values of CPS-positive and CPS-negative parameters, as well as the latest CS Score of both the teams, are displayed during CPS play only

CFE   Continuous Fuzzy Evaluation

CFE Score (CS)   Continuous Fuzzy Evaluation Score or CFE Scores or CS

CFE software   During the play, the data are stored immediately in the concerned database in the FIFA server or UEFA server for a software called by 'CFE Software' which will be executed nine times during the 90 minutes of play, once in every 10 minutes

CPS   CFE Penalty Shootout (different from the existing concept of 'Penalty Shootout')

FIFA   Fédération Internationale de Football Association

Fuzzy Pocket Machine	This is a handy small simple electronic wireless machine M looking like a mobile phone using which the Referee awards a 'punishment fuzzy set' A of F within a maximum of ten to twelve seconds at real time
Goal Post	A 'goal post' in CFE is a physical structure consisting of four components which are two vertical posts, a horizontal crossbar supporting the vertical posts and the net attached behind the goal frame
Goal-Area Post	A 'Goal-Area Post' in CFE is a physical structure consisting of four components which are two vertical posts, a horizontal crossbar supporting the vertical posts and the net attached behind the frame. The structure for a 'Goal-Area Post' is defined as a rectangular structure frame 36 feet wide by 14 feet tall that is placed at each end of the playing field. A net is attached behind the frame to catch the ball which is marginally missing to be a goal(G)
Goal Shot	A deliberate attempt by a team on the goal of his opponent team by a kick or by head or by any valid way is referred to as a 'Goal Shot'
Half-ground of the Opponent	The 'Half-ground of the Opponent for team X' at any moment is the half field which is closer to the goalkeeper of Y, and the 'Half-ground of the Opponent for team Y' is the half field which is closer to the goalkeeper of X
IFAB	International Football Association Board
MGS	By a 'Missed Goal-Shot' (MGS), we mean that the goal shot was an attempt for goal, but marginally missed the success and hence goes outside the 'Goal Post' getting into the net of the 'Goal-Area Post'. Needless to mention that the ball must cross the goal line completely. It is assumed that the goal shot corresponding to an MGS is valid as per Laws of the Game, even not invalidated due to handball, offside, etc. **Any goal shot, if qualified to be an SCB (as per definition of SCB given in Sect.** 7.2 **), will never be qualified to be an MGS even if it enters into or does not enter into the net of Goal-Area Post**
positive/negative parameters	There are 17 parameters (for each team) in the theory of CFE. Seven of them are negative parameters and rest ten of them are positive parameters. Positive parameters are those which can extract the merits of a team and negative parameters are those which can extract the demerits of a team
TCFE-1	20 minutes extra play (10 minutes + 10 minutes) and with the application of CFE during this 20 minutes of play. Or 30 minutes extra play (15 minutes + 15 minutes) and with application of CFE during this 20 minutes of play

TCFE-2	CPS play
TCFE-3	It is a sequential combination of TCFE-1 and TCFE-2. In other words, TCFE-3 means first TCFE-1 play and then TCFE-2 play will take place
UEFA	Union of European Football Associations

Abstract

This book presents a proposal for a new soft computing innovative method called by **CFE** for evaluation of the football matches of FIFA (IFAB) and UEFA/EFL to compute the true 'Winner'. It is a research work in the area of 'Football Science', being proposed to FIFA (IFAB) and UEFA/EFL for consideration to replace the existing obsolete and weak rules of football matches of 90 minutes to declare the final 'Winner'. In CFE, the final decision about any football match of 90 minutes is obtained 'by computation', by execution of a software called by 'CFE software' with the real-time input values directly from the field on a number of highly significant parameters, some of them being continuous variables. The result of m-n goals at the end of 90 minutes play in a football match is just one of the many parameters for the computation of the 'Winner' in our proposed CFE method. The evaluation is mainly based upon the application of fuzzy logic, 'Fuzzy Pocket Machine', fast computer server, good communication system from playing field to the server (as used in cricket matches) and the CFE software at the server. The referees need not be experts in fuzzy set theory; they can be easily trained within just 30 minutes of demonstration on how to use the 'fuzzy pocket machine' for giving real-time inputs to the database at the server.

The existing method of FIFA/UEFA/EFL chooses the 'Winner' team directly on the basis of one and only one parameter which is the 'm-n goal score' at the end of 90 minutes of play, ignoring many other important and significant continuous parameters of the 90 minutes duration since the start of the game. The existing method has no scope to compute the 'better' team because of its obsolete and weak rules, because of non-availability of any excellent way in the existing rules of FIFA (IFAB) and UEFA/EFL.

Our proposed method CFE does not focus at declaring the 'Winner' directly, but at finding the actually 'BETTER' team by continuous evaluation and by computation of all the real-time data/information available from the field since the start of the 90 minutes game. Consequently, at the end of 90 minutes of play, the CFE software is executed to compute the 'better' team first of all, and then the 'better team' is finally declared to be the 'Winner' of the game, the philosophy which does not have any open or hidden chance to yield any amount of frustration to the football fans, to both the teams and their supporters, and in particular to the looser team and its supporters, as the proposed CFE method is very much sound, highly scientifical and technical, complete and logical.

The abbreviation **CFE** stands for **Continuous Fuzzy Evaluation** because CFE is basically a fuzzy constructive method by which 'CS Score' of each team is computed. It is called 'continuous' because of the reason that mathematically the 'CS Score' is a function of several variables, and for each team, the corresponding 'CS Score' varies continuously at every second, some of the parameters of this function being continuous variables of time. This proposed fuzzy method can easily extract the true dominant team out of the two teams, executing the 'CFE software' for which there will be continuous real-time inputs in a database directly from the fields and from the Referee where some of the inputs are given by the Referee using his **fuzzy pocket machine** M. The **fuzzy pocket machine** M is an electronic wireless machine (looking like a mobile phone) having its own memory and keeps backup of all the data whatever be communicated by the Referee to the database for the CFE software at the server. Besides this physical machine, the introduction of a physical structure called by 'Goal-Area Post' fixed just outside the goal line at each of the two ends is a new idea which helps to provide a good kind of input to the database at the server. It is claimed that if FIFA (IFAB) implements this method of **CFE** in World Cup football matches, it will be a true justice to the football world, to the fans, to the players, to both the teams and their supporters (in particular to the looser team and its supporters) and consequently it will give enormous justice to the 'football'. The CFE method, if implemented by FIFA (IFAB) and UEFA/EFL replacing the existing football rules, can improve the football by a huge momentum as a game and, as a subject, can do huge justice to the football fans and organizers and all concerned staff, and in particular to the looser team. The FIFA (IFAB) and UEFA/EFL can well employ the proposed machine 'Robot Referee' to replace the human Referee in football matches for continuous evaluation with very high precisions, the huge advantages of which are justified and explained in length. With the very rich and advanced theories of science and technology available in the literature in the present century, the cognition system of the machine 'Robot Referee' can be well equipped with enormous amount of artificial intelligence and knowledge as per football norms and theories. One of the major advantages of using the highly intelligent 'Robot Referee' in football matches is that its amount of intelligence and knowledge, its thinking capability and speed, its speed of physical movement on the field, etc. can be enhanced as much as required by FIFA (IFAB) and UEFA/EFL time to time, say once every 10 years, with the further development of new technologies.

The theory of CFE does not change the rules of football game (except during the very rare case of CPS play, if any), but it introduces major changes and improvements in the evaluation method and in the method of arriving at the final decision.

Keywords FIFA · IFAB · UEFA · EFL · Fuzzy Set · Half-ground of the Opponent · CFE · CFE software · Goal-Area Post · MGS · BGS Parameters · CFE Score (CS) · Fuzzy Pocket Machine · Punishment Fuzzy Set TCFE · CPS · Board-1 · Board-2 · Robot Referee

Continuous Fuzzy Evaluation Methods: A Novel Tool for the Analysis and Decision Making in Football (or Soccer) Matches

A New Innovative Proposal to FIFA & UEFA

1 Introduction

The main problem of the present football is that in <u>many</u> of the games the 'actual better team' does not become 'Winner' at the end of 90 min of the play or at the end of extra play or at the end of penalty shoot out play. In fact it is a frequent outcome in football games. This problem is identified after realizing the shocking hearts and frustration of football fans of the world lying hidden, silence and overlooked/ignored for so many years, and thus it is a long standing unsolved problem. In every FIFA World Cup, in every UEFA Euro Championship or in every big tournament, it can be observed that there happened <u>many</u> cases where the actually better team is finally happened to be declared as the 'looser' as per the existing FIFA(IFAB)/UEFA/EFL rules. The existing FIFA(IFAB)/UEFA/EFL rules does not have any scope to judge the better team by any means of science and technology, but only to wait for one and only one data at the end of 90 min of play for declaring the 'Winner' or for going for 'Extra Play' and/or 'Penalty Shootout', etc. to find one team whom to be declared as the 'Winner'.

For instance [26–29], consider the recent **UEFA Euro 2016** held in France from 10th June to 10th July 2016. On 10th July'2016, the Portuguese beat host France 1-0 in the final played at the Stade de France in Paris, managing a sudden extra-time goal, although during 90 min of play France showed much superior quality of football elements. Just at the end of this game, the emotional shocking outburst of France supporters at Paris roads are known to the world!

In this Championship the greatest surprise is due to the fact that during their seven games in the tournament they managed to win just one match inside 90 min!. Portugal progressed from a weak looking Group F in third place after drawing their games with each of Hungary, Iceland and Austria.

© The Author(s) 2018
R. Biswas, *Continuous Fuzzy Evaluation Methods: A Novel Tool for the Analysis and Decision Making in Football (or Soccer) Matches*, SpringerBriefs in Computational Intelligence, https://doi.org/10.1007/978-3-319-70751-8_1

In the Round of 16, due to a tie of Portugal against Croatia the game went to extra-time after ending 0-0 before Ricardo Quaresma scored a winner in the 117th minute. In the next round of Portugal against Poland, penalties were needed after the game ended 1-1. The semi-final against Wales was the only game Portugal managed to win inside 90 min.

And at the Stade de France in Paris against the host France the match, all too predictably, wasn't settled in normal time. Besides all the draw cases within 90 min by Portuguese, there are many other draw cases too in each of: the Round of 16, the Quarter Final, the Semi Final, and the Final! of UEFA Euro 2016. In every UEFA Euro or in every World Cup or even in every football championship held in this world, this type of tie-cases are very common in the history. For example, scoreless draw of Brazil v Mexico in World Cup 2014 in First Stage Group-A at Estadio Castelao Fortaleza (Brazil) on 17th June'2014, Germany v Ghana 2-2 draw in World Cup 2014 in First Stage Group-G at Estadio Castelao Fortaleza (Brazil) on 21st June'2014, etc. to list out of many.

Such shocking decisions of the football games is happening:

(i)   **NOT** due to any fault of the winner team or looser team,
(ii)   **NOT** due to any fault of the FIFA or UEFA or EFL officials,
(iii)   **NOT** due to any fault of the Referees,
(iv)   **NOT** due to any fault of the fans/crowds,
(v)   **NOT** due to any fault of the supporters,
(vi)   **NOT** due to any fault of the stadium/field,
(vii)   **NOT** due to any fault of the weather condition, etc.

but due to the very obsolete and weak rules of FIFA(IFAB).

The existing set of FIFA rules was **not weak** even 30–40 years back, but today in this century it is surely so. In the last decades, the subjects like: Mathematics, Statistics, Soft-computing, Computation Techniques, Decision Theory, Judgment Methods, Computer Engineering, Electronics Engineering, Mechatronics Engineering, Communication Engineering, etc. have improved by a huge momentum by the significant amount of rich research work carried out by the world scientists in Sciences and Technologies. In particular, the areas like Soft-computing, Fuzzy Logic, Communication/Instrumentation, Information & Communication Technology (ICT), Computer Science, Software Engineering, Mechanical Engineering, etc. have changed the computation techniques of the present world to a very very great height. But FIFA(IFAB) could not use and apply the latest development of science and technology to improve or to update its rules for football matches, in particular its evaluation process to find out the 'Winner' at the end.

In this research work the author proposes a highly improved method to FIFA (IFAB) and UEFA/EFL on how to select the 'Winner' of a football match which is a kind of continuous soft-computing evaluation method. The method is called by

CFE which is basically a Fuzzy constructive method. The abbreviation **CFE** stands for the phrase "Continuous Fuzzy Evaluation", the method which mainly employs the powerful soft-computing mathematical tool 'Fuzzy Logic', Computer Server, Fuzzy Pocket Machine, and ICT. It is called to be a 'continuous' evaluation method because of the reason that mathematically the 'CS-score' of each team varies continuously at every second, some of the parameters of the function 'CS-score' being continuous variables of time. This proposed fuzzy method can easily extract the true dominant team out of the two teams immediately after the 90 min of play using a software called by 'CFE-software', in the database of which continuous real time inputs reach directly from the fields and from the Referee. Many of the inputs given by the Referee are communicated to the server using his **'fuzzy pocket machine'** M which is an electronic wireless machine (looking like a mobile phone). The **'fuzzy pocket machine'** M has its own memory too. The CFE computes the 'better team' at the server by continuous data of 90 min available from the playing-field, by executing the CFE-software, in which one of the many important parameters used is the 'm-n goals' score at the end.

The breakthrough in the Theory of CFE is that the 'm-n goals' result at the end of 90 min of play does not necessarily mean that the team scoring m goals is the winner, even if m > n.

The breakthrough in the Theory of CFE is also due to the fact that the 'm-n goals' result at the end of 90 min of play does not necessarily mean that it is a tie/draw case, even if m = n.

For a football match the CFE-software computes who is the 'actually better' team, and then the 'actually better' team is officially declared to be the **'Winner'**. In our article here, we frequently use the synonyms 'CFE play' or 'CFE game' which means that the concerned football game is played under the norms of CFE method.

It is claimed that if FIFA (IFAB) and UEFA/EFL implement this method of **CFE** in World Cup Football matches or Euro Cup Football matches, etc. then it will surely provide a true justice and enormous amount of satisfaction to the football world, to the fans, to the players, to both the teams and their supporters (in particular to the looser team and its supporters) and consequently it will give enormous justice to the 'football' if considered as a subject.

During the last five decades, the Fuzzy Theory [13–25] discovered in 1965 by Prof. Zadeh [24] has been fluently applied in various problems in almost all areas of Science, Technology, Social Science, Law, Medical Science, etc. to list a few only out of many. But most probably no work has been so far reported in literature showing the application of fuzzy theory in the domain of "Sports Science", and in particular in most popular sports like Football, Cricket, Tennis, Badminton, Chess, etc. Collection of statistical data and the activities for statistical analysis are commonly practiced in almost all areas of sports. Football is probably the most popular sports in the world, and every football game during every minute of its play generates continuously the real time data for statistical analysis and conclusions. In every sport, the Referees are not a simple decision makers but supposed to be the truly experts, the best available decision makers in the respective area of sports, and are always carefully recruited by the concerned sports authority.

Important Presumption:

Consequently, in football games too, we have to presume the following:

(i) there is '**no hesitation part**' corresponding to every resultant-decision taken by the Referees during the period of play,
(ii) Referees' decisions are final.

For such type of problems, where decision makers are most intellectual and knowledgeable on the concerned subject, where decision makers are pre-selected by top experts as genuine, excellent and extra-ordinary decision makers, it is surely the 'Fuzzy Theory' to be regarded as the most appropriate tool to deal with, even compared to the more powerful [9, 10] soft-computing tool 'Intuitionistic Fuzzy Theory' of Atanassov [1–8].

2 'Penalty Shootout' or 'Extra Play': Not Required in CFE Method

The '**penalty shootout**' is a method presently being followed in FIFA/UEFA matches for determining a winner in Football matches that would have otherwise been drawn or tied after 90 min of play [26–29]. The penalty shootout is normally used only in situations where a Winner is needed to be identified and where other methods such as **extra time** and **sudden death** have failed to determine a winner. It avoids the delays involved in staging a **replayed match** in order to produce a decisive result. The term **golden goal** was introduced by FIFA [26–29] in 1993 along with the rule change because the alternative term 'sudden death' was perceived to have negative connotations. Following a draw, two fifteen-minute periods of extra time are played as per the existing rules [26–29]. If any team scores a goal during extra time, that team becomes the winner and the game ends at once. The winning goal is known as the 'golden goal'. If there have been no goals scored after both periods of extra time, a penalty shootout decides the game. These are very popular rules in FIFA/UEFA and are being followed immediately while the situation arises. The first ever 'penalty shootout' adopted in the World Cup was on 9th January 1977, in the first round of African qualifying, when Tunisia beat Morocco. The first ever 'penalty shootout' in the finals tournament was in 1982, when West Germany beat France in the semifinal.

In our **Theory of CFE** we regard very seriously the fact that the 'penalty shootout' is a test of few individuals which may be considered inappropriate in a team sport, in particular where a team size is large like 11 (eleven)!. Football is a 'team sport' and penalties is not a team, it is the sport-element by individual. The most interesting and important merit of a football game is that the game is played for 90 min with a complete mutual and continuous understanding, continuous mutual coding/decoding of the 11 players. It is played by a continuous performance of the 11 players together, every second is countable and accountable to them during the 90 min of play.

But it is shocking fact that the inferior teams are tempted sometimes to play preferably for a scoreless draw or for any n-n draw, calculating that a shoot-out may offer their best hope of victory. Killing play-time <u>as maximum as possible</u> **during 90 min of play is not an impossible job or tough job of a team within the existing rules of FIFA(IFAB) or UEFA.** The FIFA/UEFA does not have any official way to deal with such real but boring situations happening frequently in the matches being witnessed by the frustrated fans. The FIFA(IFAB)/UEFA is really incapable to do sufficient justice to the football in such type of real and frequent cases.

For a unique example, consider the 1990 FIFA World Cup which was notable for many teams playing defensive football and using time wasting tactics, including even a team of very high caliber who scored only 5 goals but reached the final by winning two shootouts. As a way to decide a football match, shoot-outs have been seen variously as a thrilling climax or as an unsatisfactory cop-out. The result is often seen as an **exciting lottery** rather than a test of sports-skill. Only a small subset of a footballer's skills is tested by a penalty shootout, not the skill of the team as a whole.

One of the very rich and strong merits of our proposed CFE method is that it <u>does not need</u> 'penalty shootout' for deciding the Winner of a football match that has ended with n-n goals after 90 min of play, and does not need even any 'extra play'.

However, the notion of **CPS** play introduced in Sect. 19 in this article here is a <u>different</u> concept with a very beautiful philosophy and justice. The **CPS play** in our proposed CFE method and the '**penalty shootout**' in the existing FIFA method are not same. But in reality in almost all the cases evaluated by CFE, the CPS is not required in CFE, probably not even once in 1000 played games!, because of its very very low chance of requirement.

3   Brief Introduction of Fuzzy Set Theory

The Theory of CFE is a soft-computing based theory developed here using fuzzy set theory. Therefore a very brief introduction of fuzzy set theory is presented here for the readers for ready reference. In [9, 10] it is justified in length that the present world is growing not just by human being and their cognitive domain alone, but on the basis of continuous type of Human-Computer interaction that is concerned with the joint execution of tasks by human and machines; the architectural structure of communication between human and machine; human capabilities to use machines, algorithms and programming of the interface itself. Human-computer interaction thus has a recall for science, engineering, and design aspects using a more natural logic than the existing crisp logic as justified mathematically in [9, 10]. Two-valued classical logic basically forms a well-known boundary case of the fuzzy logic.

Prof. Zadeh [24] in 1965 initiated the notion of fuzzy set theory as an extension of the ordinary set theory, which turned out to be of far reaching implications.

Imprecise or vague notions can be well modeled using this theory. A fuzzy set is a class of objects in which the transition form membership to non-membership is gradual rather that abrupt. Such a class is characterized by a membership function which assigns to an element a grade or degree of membership between 0 and 1, both inclusive.

3.1 Fuzzy Set

Fuzzy sets are often defined through membership functions to the effect that every element is allotted a corresponding grade of membership from the unit interval [0,1] in the fuzzy set. Consider a fuzzy set C of the universe X. The membership function μ_c that determines the grades of membership of individual elements x in the fuzzy set C must satisfy the following constraint:

$$0 \le \mu_c(x) \le 1 \quad \forall x \in X.$$

In the real world the human reasoning in most of the cases involves the use of variables whose values are fuzzy sets, besides crisp variables. Description of system behavior in the language of fuzzy theory lowers the need for precision in data gathering and data manipulation. Essentially, in a fuzzy set each element is associated with a point-value 'membership value' selected from the unit interval [0,1], which is also termed the 'degree of belongingness' or 'grade of membership' of the element in the fuzzy set.

Mathematically, if X be a universe of discourse, a **fuzzy set** A in X is a set of ordered pairs

$$A = \{(x, \mu_A(x)) : x \in X\}$$

where $\mu_A \colon X \to [0,1]$ is a function called by "**membership function**" of the fuzzy set A. Thus, $0 \le \mu_A(x) \le 1 \quad \forall x \in X$.

Here X is the universe of discourse, or the universal set, which contains all the possible elements of the concern problem which is under study and investigation. The membership function μ_A maps each element of X to a membership grade (or membership value) between 0 and 1, being proposed by the concerned decision maker.

3.1.1 α-Cut of a Fuzzy Set

Consider a **decision parameter** $\alpha \in [0, 1]$. An **α-cut** of a fuzzy set A is a crisp set A_α that contains the elements of X that have membership value in A greater than or equal to α. i.e. $A_\alpha = \{x \mid A(x) \ge \alpha\}$. Thus $A_\alpha \subseteq X$. Also $A_0 = X$ for $\alpha = 0$.

A **strong α-cut** of a fuzzy set A is a crisp set $A_{\alpha+}$ that contains the elements of X that have membership value in A strictly greater than α. i.e. $A_{\alpha+} = \{x \mid A(x) > \alpha\}$. Thus $A_{\alpha+} \subseteq A_{\alpha} \subseteq X$. Clearly $A_{1+} = \varphi$ (null set).

3.1.2 Various Operations on Fuzzy Sets

In this section we recollect some basic operations on fuzzy sets. Let A and B be two fuzzy sets of the universe X having membership functions μ_A and μ_B respectively.

Equality of two fuzzy sets
Two fuzzy sets A and B are equal if and only if $\mu_A(x) = \mu_B(x)$ $\forall x \in X$.

If $\exists$ at least one $x \in X$ such that $\mu_A(x) \neq \mu_B(x)$, then A and B are said to be 'not equal' and it is denoted as $A \neq B$.

Fuzzy Subset
Fuzzy set A is a subset of the fuzzy set B if and only if $\mu_A(x) \leq \mu_B(x)$ $\forall x \in X$.

In other words, $A \subset B$ if for every x of X the grade of membership in fuzzy set A is less than or equal to the grade of membership in fuzzy set B.

For a mathematical example, let X = {a, b, c, d} be the universe, A = {a/0.9, b/0.2, c/0.1, d/0.6} and B = {a/0.6, b/0.1, c/0, d/0.1} are two fuzzy sets of X. Then, clearly $B \subseteq A$.

For another example, let us denote by A and B, respectively, the sets of "long" and "very long" travel times. The fuzzy set "very long" travel time is a subset of the fuzzy set "long" travel time since the following relation is satisfied for every x:

$$\mu_B(x) \leq \mu_A(x).$$

Complement of a Fuzzy Set
The fuzzy set A is called to be the complement of the fuzzy set B i.e. $A = B^c$ if

$$\mu_A(x) \quad \text{i.e.} \ \mu_{Bc}(x) = 1 - \mu_B(x), \quad \forall x \in X.$$

Clearly, we have $(B^c)^c = B$.

For example, let X = {a, b, c, d} be the universe, and A = {a/0.2, b/0.6, c/0.3, d/0.1}, B = {a/0.8, b/0.4, c/0.7, d/0.9} are two fuzzy sets of X. Then clearly $A = B^c$ or equivalently $B = A^c$.

Union of two Fuzzy Sets
The union of two fuzzy sets A and B is denoted by $A \cup B$ and is defined as the smallest fuzzy set that contains both the fuzzy sets A and B.

The membership function $\mu_{A \cup B}$ of the union $A \cup B$ of the two fuzzy sets A and B is defined as follows:

$$\mu_{A \cup B}(x) = \max\{\mu_A(x), \mu_B(x)\} \quad \forall x \in X.$$

The union corresponds to the operation "OR".

Example:

For example, let X = {a, b, c, d} be the universe, A = {a/0.3, b/0.2, c/0.4, d/0.7} and B = {a/0.2, b/0.9, c/0.7, d/0.7} are two fuzzy sets of the universe X.

Then $A \cup B$ = {a/0.3, b/0.9, c/0.7, d/0.7}.

Intersection of Fuzzy Sets

The intersection of two fuzzy sets A and B is denoted by A ∩ B and is defined as the largest fuzzy set that is contained in both the fuzzy sets A and B. The intersection corresponds to the operation "AND".

Membership function $\mu_{A \cap B}(x)$ of the intersection A ∩ B is defined as follows:

$$\mu_{A \cap B}(x) = \min\{\mu_A(x), \mu_B(x)\quad \forall x \in \mathbf{X}.$$

Example:

For example, let X = {a, b, c, d} be the universe, A = {a/0.3, b/0.2, c/0.4, d/0.7} and B = {a/0.2, b/0.9, c/0.7, d/0.7} are two fuzzy sets of the universe X.

Then $A \cap B$ = {a/0.2, b/0.2, c/0.4, d/0.7}.

3.2　Intuitionistic Fuzzy Set (IFS)

Different authors from time to time have made a number of generalizations of the fuzzy set theory of Zadeh [24] and in parallel developed alternative soft-computing set theories to deal with the real life imprecise problems. The notion of intuitionistic fuzzy set theory (IFS theory) introduced by Atanassov [1–8] is a very useful and probably the richest and most potential soft-computing set theory ever discovered, as justified in details in [9, 10]. All fuzzy sets can be viewed as intuitionistic fuzzy sets, but the converse is not true. There are a large number of real life problems for which IFS theory is a more suitable tool than fuzzy set theory.

In most cases of judgments, evaluation is done by human beings where there is a limitation of knowledge or intellectual capabilities. There is no doubt in it. Naturally, every decision-maker hesitates more or less, on every evaluation activity. It is a common feature of any human being. To decide "whether 5 + 7 = 12 or 'not 12'", the hesitation is finally nil to the decision maker. But to judge whether a patient has cancer or not, a doctor, (the decision-maker) will hesitate because of the fact that a fraction of evaluation may remain indeterministic to him. This is the breaking philosophy in the notion of IFS theory introduced by Atanassov. The non-membership part may have more significant importance compared to the 'complement of fuzzy sets'. If there is no hesitation, the intuitionistic-fuzziness reduces to fuzziness and in such a case there is absolutely no need to exploit intuitionistic fuzzy tools for analyzing the problem. Atanassov introduced the notion of intuitionistic fuzzy sets by treating membership as a fuzzy logical value rather than a single truth value.

If X be a universe of discourse, an intuitionistic fuzzy set A in X is a set of ordered triplets

$$A = \{\langle x, \mu_A(x), \upsilon_A(x) \rangle : x \in X\}$$

where $\mu_A: X \to [0,1]$ and $\upsilon_A: X \to [0,1]$ are functions called by "membership function" and "non-membership function" respectively such that

$$0 \leq \mu_A(x) + \upsilon_A(x) \leq 1 \quad \forall x \in X.$$

For each $x \in X$, the values $\mu_A(x)$ and $\upsilon_A(x)$ represent the "degree of membership" (or membership value) and "degree of non-membership" of the element x to the intuitionistic fuzzy set (IFS) A respectively.

And the rest amount $\pi_A(x)$ given by

$$\pi_A(x) = 1 - \mu_A(x) - \upsilon_A(x)$$

is called the "indeterministic part" or "hesitation part". The value of $\pi_A(x)$ is also called the degree of non-determinacy (or uncertainty) of the element $x \in X$ to the intuitionistic fuzzy set A.

We call the inequality condition $0 \leq \mu_A(x) + v_A(x) \leq 1$ by "Atanassov condition".

Of course, a fuzzy set may be written as

$$\{(x, \mu_A(x), 1 - \mu_A(x)) | x \in X\}.$$

Thus, a fuzzy set is a particular case of the intuitionistic fuzzy set if $\pi_A(x) = 0, \forall x \in X$, i.e. every fuzzy set is an intuitionistic fuzzy set but not conversely.

For details of the classical notion of intuitionistic fuzzy set (IFS) theory, one could see the books authored by Atanassov [1–8]. A detailed analysis of intuitionistic fuzzy set theory as a very powerful and appropriate soft-computing theory for almost all the ill-defined problems has been done by Biswas in [9, 10].

But it is justified earlier in the last two paragraphs in Sect. 1 of this article that to develop our Theory of CFE, the fuzzy theory will be an appropriate soft-computing tool than any other. The application of Intuitionistic fuzzy sets instead of fuzzy sets means the introduction of another degree of freedom into a set description and consequently the complexity will be more in this case.

4   Introducing the "Theory of CFE" for FIFA Matches

In this article we introduce a new theory for football science called by the "**Theory of CFE**" to the FIFA/UEFA/EFL authority. In this new theory we basically present a new innovative method called by the '**CFE Method**', introducing several new terms, rules and related literature. The Theory of CFE does not change the FIFA rules of football game (except during CPS Play, if any), but it introduces a major changes and huge improvements in the evaluation method of a football game and in the method of arriving at the final decision about 'Who is the Winner'. The Theory of CFE does not also violate/contradict the FIFA rules of football game.

Our proposed method of **CFE** (**C**ontinuous **F**uzzy **E**valuation) is applicable to all the football matches of 90 min of play (or, to any **t** minutes of play). This method is developed to replace the existing very <u>obsolete</u> and weak method presently followed by FIFA and UEFA/EFL. The CFE is a multi-criteria based decision making process to declare the 'Winner' by computational method processed at the server by executing a software called by 'CFE-software'. A team's performance, its merits and demerits, are computed at any time during and at the end of the 90 min game on the basis of certain significant parameters, not on the basis of just one and only one parameter which is the 'm-n goal score' at the end of 90 min of play.

It is explained earlier in Sect. 2 that another major weakness in the existing method of FIFA/UEFA/EFL is that in case of tie a penalty shootout is used to decide <u>somehow</u> a 'Winner' of a football match, where the other methods such as extra time and/or sudden death have failed to determine a winner. The penalty shootout method has a lot of demerits, yields a lot of frustration. It does not give satisfaction to the looser, and even not to the winner sometimes, and surely not to the fans who have watched the game critically for 90 min. The spirit of the sport is not translated to the result, instead it becomes equivalent to a <u>lottery game</u> in many situations. The probability that the better team will win is not high, because in many cases it becomes like that: "the better goalkeeper will win, not necessarily the better team". This is a genuine unsolved long-standing problem to the world football fans and teams and players, and thus to the football sport if considered as a subject of study and research.

In our CFE method, there is no extra time of play, there is no penalty shoot out. All the games can be decided by 90 min of play except in very very few cases (possibility of which is too low) which may not even occur once in 1000 games!.

The proposed method "CFE" in fact does not focus at finding the 'Winner' directly, but at finding the actually 'better' team by <u>continuous</u> evaluation and by computation of all the real time data/information available since the start of the game. Consequently, at the end of computation the 'better' team is declared to be the 'Winner' of the game. This philosophy does not carry any frustration and is very much sound, complete and logical. This philosophy will neither hurt the hearts of the football fans nor will give any amount of injustice to the looser team, unlike the existing situation in FIFA and UEFA/EFL games.

For a football game, we therefore propose the most important problem which is:

"How to choose the 'better team' when the number of goals scored in this match are 'm-n goals' at the end of 90 min of play, where either m = n or m ≠ n?".

The existing FIFA method chooses the 'Winner' team directly, without computing the 'better' team because of its weak and obsolete rules. The method just checks whether the condition m $\neq$ n is true or not, and if true then immediately the Winner is declared. The decision is based on the basis of one and only one data which is the 'm-n goals' at the end of 90 min play. This can not (should not) be an appropriate approach in this century, as today we are having a rich volume of literatures of research work done by the world scientists in Science and Engineering, in particular in Mathematics, Statistics, Soft-computing, Computation Techniques, Decision Theory, Judgment Methods, Computer Engineering, Electronics Engineering, Mechatronics Engineering, Communication Engineering, etc. Consequently, it is fact that we must develop a new innovative method/rules so that the better team should win, not the weaker team, **by any chance**. It is observed in several cases in every World Cup or EURO Cup matches that the existing method of FIFA(IFAB) and UEFA/EFL can not guarantee it.

The "**Winner**" between the two teams:

(i)   must be the overall <u>dominating</u> team of 90 min, not a suddenly dominating team for few minutes by scoring a goal (for example, say suddenly scoring 1-0 goal).

(ii) must be the overall <u>better</u> team by continuous performance shown by the team, who has contributed more substance of the subject 'football' in today's game by way of quality and performance during the 90 min of play.

The existing FIFA method of taking decision on the basis of the results of 'm-n goals' after 90 min of play to declare the 'Winner' is a weak method, because it can not comply with the above two conditions.

During the play of a football match, each team plays and performs by its best skill revealing its merits in various criteria/attributes continuously at every moment. But each team also happens to commit mistakes during play. Out of which some mistakes are of major nature not committed continuously but at discrete moments of time. For example, a team commits one foul after 23 min (without Yellow card or Red card), another foul after 12 min (without Yellow card or Red card), then another foul but this time shown yellow card to one player of this team by the referee after 18 min, etc.

These events if quantified by good mathematical techniques (using hybrid of soft-computing and hard-computing methods) during the 90 min span of play can surely indicate the amount of discredits, amount of poor performance etc. of each of the teams. These events are recorded by the organizer for statistical analysis only. In our proposed CFE method we input such discrete negative events by extracting data/information in a continuous manner during the 90 min of play, in addition to the positive events. Finally we compute the individual **Continuous Fuzzy Evaluation Score** or **CFE-Score** or **CS** (in short) of both the teams by executing the CFE-software at the server, comparing which we understand which team is the truly better performer today, i.e. which team is the truly dominant team today. The individual CS-Score for each team can also be computed at any time during the 90 min of play to observe the latest status of performance of them. The team with higher CFE-Score (CS) computed at the end of 90 min play will be declared as the **Winner**.

4.1 "Half-Ground of the Opponent" for a Team in CFE

To introduce the **Theory of CFE** we first of all define the term "**Half-ground of the Opponent**" at any instant of time for each of the two teams by a figure here. Consider a football match to be played between two teams X and Y under CFE method of evaluation. The Fig. 1 shows that team X is on the left hand side and team Y is on the right hand side during one half of the match. For the team X, the land "Half-ground of the Opponent" is shown by arrow mark. Similarly, for the team Y, the land "Half-ground of the Opponent" is also shown by arrow mark. These two half-grounds is the partition of the complete ground into two equal halves by the middle half-line.

Thus the "Half-ground of the Opponent for team X" at any moment is the half field which is closer to the goalkeeper of Y, and the "Half-ground of the Opponent for team Y" is the half field which is closer to the goalkeeper of X. Obviously, after the play of the first half, while the teams change the goal-posts, the "Half-ground of the Opponent" will also be changed for each team.

4.2 Basic Rules in "CFE" (for Almost All the Games)

The method of CFE is based on computation, and is well applicable to any football match of 90 min. It is claimed that CFE method being so rich both scientifically and technically can compute the 'better' team easily by executing the CFE-software with the continuous real time input data available during the 90 min of play. However mathematically there exist a very very low possibility that CFE depends upon its CPS play (see Sect. 19 for details about CPS play in CFE), in reality such a case may not occur even once in 1000 played games.

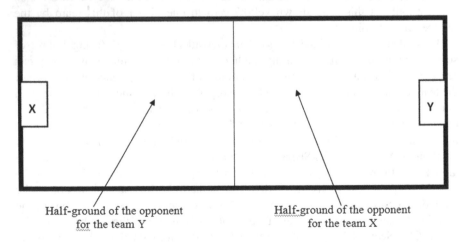

Half-ground of the opponent   Half-ground of the opponent
      for the team Y   for the team X

Fig. 1 Two 'Half-ground of the Opponent': one for the team X and the other for the team Y

In "CFE" the following basic rules are applicable, in any general case and <u>in almost all the cases</u> (ignoring the very very rare case whose possibility is not just zero mathematically while CPS play may be required to be played, the notion of CPS is explained later in this article in Sect. 19):

(i)   There is **no** 'extra time of play' beyond 90 min.

(ii) There is **no** 'Replay' beyond 90 min.

(iii) There is **no** 'Penalty Shootout'.

(iv) There is **no** 'CPS', in almost all the CFE cases (except very very rare situations). (The notion of CPS is explained later in this article in Sect. 19)

(v)   There is **no** 'sudden death' round.

(vi) Every game can be well decided within 90 min of CFE play (except very very rare situations).

(vii) In CFE, a result of m-n goals after 90 min of play, where m > n, does not <u>necessarily</u> mean that the team scoring m number of goals is the <u>better</u> (winner) team.

(viii) A result of n-n goals at the end of 90 min play does not necessarily mean that it is a tie/draw case, because in CFE method a result of n-n goals does not necessarily mean that 'there is no scope to determine the <u>better</u> team' without a further amount of play. This is the breakthrough path in the CFE method.

(**Note**: During the last 50 years or so, a lot of innovative methods of scientific and technological computation have been discovered by the world scientists. Consequently, a large number of problems (both soft-problems and hard-problems) which were having a poor quality of solutions in earlier century can now be solved more accurately, more precisely, with much higher amount of satisfaction. Our proposed CFE is a much improved method, both scientifically and technical, compared to the existing obsolete and poorly ruled method followed by FIFA(IFAB) and UEFA/EFL.)

(ix) **The method of "CFE" in fact does not focus at finding the 'Winner' <u>directly</u>, but at finding out the actually 'BETTER' team by <u>continuous</u> evaluation and computation of all the real time data/information available since the start of the game**.

Consequently, at the end of 90 min of play, the computed 'better' team is declared to be the 'Winner' of the game, the philosophy which does not carry any contradiction and is very much sound, complete and logical. This philosophy will neither hurt the hearts of football fans nor will give any amount of injustice to the football game as a subject, unlike the present situations in most of such cases. This philosophy enrich the method of "CFE" by a huge potential and strength, by a huge amount of justice to the football compared to the existing rules followed by FIFA(IFAB) and UEFA/EFL.

(x)   The "CFE" evaluation method for the final decision of the game is an <u>integrated</u> computing approach of a number of real time data generated during the <u>continuous</u> inspection and refereeing since the start of the game for the complete duration of 90 min. Data of every moment of several parameters are taken into account. The final decision is not given on the basis of just one and only one piece of data which is the 'm-n goal score' data at the end of the 90 min of play. All the data (including the last piece of data which is the m-n goal score data at the end of the 90 min of play) are input to the CFE-software, and hence are important for final computation and judgment. Thus the CFE procedure is based on the 'human machine interaction', Fuzzy Logic and the execution of CFE-software in the fast computer (FIFA/UEFA server) with all the inputs of 90 min where some of the inputs are communicated from the 'fuzzy pocket machine' of the Referee (see Sect. 9). The input parameters and their domain values are explained precisely in this article in the next sections.

(xi)   The CFE-software will be executed nine(9) times during 90 min of play, once in every 10 min, by the real time data whatever so far been input to the server. Each of these nine executions provides the latest updated CS-score of each team which are displayed at the 'Electronic Display Board' for information to the football fans watching the game inside the stadium and to the world fans watching the game in TV outside the stadium. Obviously, the final decision will be given on the basis of the last execution of CFE-software i.e. the 9th execution at the end of the 90 min of play. At each of these nine times display of CS score, the 'Electronic Display Board' will also display the latest real time updated values of all the parameters (which are introduced in Sect. 7 in this article).

This above set of eleven points makes CFE method at a very strong deviation from the existing FIFA(IFAB) rules or UEFA/EFL rules. It is because of the reason that the final judgment in CFE is made on the basis of continuous data/information of complete 90 min, not by the one and only one piece of data which is the 'm-n goal' scores of the end moment. The real time continuous data/information of every moment is important in the computerized computation in CFE method for computing the better team. And finally the better team will be officially declared to be the 'Winner'.

5   Introducing the Terms: 'Goal-Area Post', Goal-Shot, G, MGS and BGS, in the "Theory of CFE"

To proceed further into the Theory of CFE, we introduce now few important terms and their rich significance in the context of every football game play. We know that the goal area is also known colloquially as the "goal box", or "6-yard box". Its purpose is to delimit the area where a goal kick is taken. A goal line marked on the playing surface between the goal posts demarcates the goal area. A goal kick in association football is the way to restart the game when the ball has been kicked past the goal line and outside the goal by the attacking team.

5.1 What is Goal-Shot?

A deliberate attempt by an attacking team on the goal of his opponent team by a **kick** or by **head** or by any valid way is referred to as a "Goal-Shot". To score a goal, the ball must pass completely over the goal line between the goal posts and under the crossbar, and no rules may be violated on the play. Thus it is our basic assumption that every goal-shot is valid as per Laws of the Game, even not invalidated due to handballl, offside, etc. In case any attempted shot violates football rules, then we will never call it to be a goal-shot.

We divide Goal-Shots (which cause the game to re-start) into three categories in our Theory of CFE:

(i)   Goal **(G)**   (scored successfully by a goal-shot)
(ii) Missed Goal-Shot **(MGS)**
(iii) Bad Goal-Shot **(BGS)**

It is important to note that a goal-shot will be one of the above three categories, but can never qualify to be fallen in two or all three categories at a time.

In our Theory of CFE, althrough in this article, by a 'Goal Shot' we will assume that it is a kick or a head or by any valid way for attempt for a goal without violating any rule of play.

5.2 "Goal Post" in CFE

The classical concept of 'goal post' according to the Laws of the Game of FIFA is known to us. The goal structure (see Fig. 2) is defined as a rectangular structure frame 24 ft wide by 8 ft tall that is placed at each end of the playing field. In most organized levels of play a net is attached behind the goal frame to catch the ball and indicate that a goal has been scored.

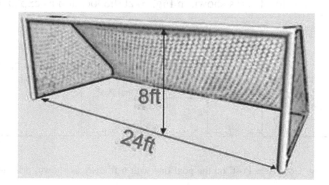

Fig. 2 Dimension of a football 'Goal Post' (collected from: https://en.wikipedia.org/wiki/Goal_ (sport))

In our Theory of CFE the concept of "Goal Post" is exactly same as the classical concept, but redefined as below:

A 'goal post' in CFE is a physical structure consisting of three components which are: two vertical posts at a distance of 24 ft, a horizontal crossbar supporting the vertical posts at a height of 8 ft.

There may be a net attached behind the goal frame.

5.3   Introducing the New Term: 'Goal-Area Post' in CFE

In Sect. 5.2 above, the notion of 'Goal Post' in CFE is reproduced. The notion of 'Goal-Area Post' is different from that of 'Goal Post'. It does not exist in the existing FIFA rules/norms of IFAB, or in any football rule in the world. It is introduced in the Theory of CFE to support the input of few data for the CFE-software.

A 'Goal-Area Post' in CFE is a physical structure (see Fig. 4) consisting of three components which are: two vertical posts, a horizontal crossbar supporting the vertical posts.

However a net may also be attached behind the frame.

The difference between 'Goal-Area Post' and 'Goal Post' is in their respective locations and dimensions. The structure for a 'Goal-Area Post' is defined as a rectangular structure frame 36 ft wide by 14 ft tall that is placed at each end of the playing field (see Fig. 5 showing the 'Goal-Area Post' at one end of the field). A net is attached behind the frame to catch the ball which is marginally missing to be a goal(G).

5.3.1   Exact Physical Location of the 'Goal-Area Post'

To understand the exact location, first of all tentatively just imagine that the 'Goal-Area Post' is located (as shown in Fig. 3) at the location as explained below:

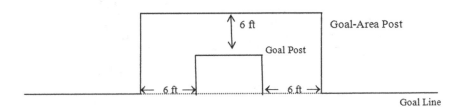

Fig. 3 Imaginary 'Goal-Area Post' on the goal line with 6 ft away as shown, prior to shifting at its exact location

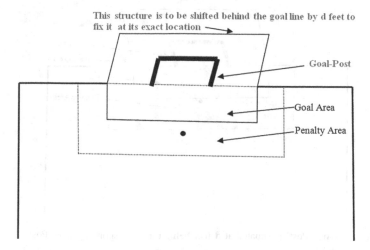

Fig. 4 A 'Goal-Area Post' is not situated on the goal line

Let us tentatively assume that the vertical posts of the 'Goal-Area Post' are situated on the goal line at the end points of the 'Goal Area' as shown in Fig. 4. The horizontal crossbar of the 'Goal-Area Post' supporting its vertical posts is situated six feet above the horizontal crossbar of the 'Goal-Post'. The vertical posts of the 'Goal-Area Post' are six feet away from their respective nearest vertical posts of the 'Goal-Post'.

For the <u>exact location</u>, now shift this tentative 'Goal-Area Post' **d** feet behind the goal line in parallel to the goal line, where **d** is at least the diameter of the ball (as prescribed by the FIFA Laws of the Game) but **d** could be at most equal to 9 in. The Fig. 5 shows the exact location of a 'Goal-Area Post'.

As per FIFA Laws of the Game, the ball is to be of a circumference of not more than 28 in and not less than 27 in. Consequently, by arithmetic calculation it appears that the value of **d** in inch unit is to be such that $8.6 \leq \mathbf{d} \leq 9$, in the Theory of CFE. This numerical value of **d** ensures that a shot can never become a MGS or BGS (see Sects. 5.4 and 5.5 below) if it is completely not outside the goal line in space (or on land).

<u>Thus, from the Fig. 5 it is clear that the both the 'Goal-Area Posts' are situated basically **outside** the playing field at the two ends.</u>

However, the various measures of the 'Goal-Area Post' can be re-fixed by the FIFA(IFAB)/UEFA experts. **If the ball touches(collides) the 'Goal-Area Post' then the ball is immediately put at 'out of play' status, even if the ball remains on the playing field. The available advanced sensor technologies may be used to incorporate the event of such kind of touch/collision.**

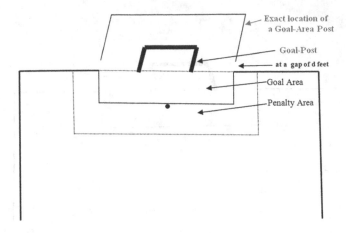

Fig. 5 A 'Goal-Area Post' is situated at **d** feet behind its corresponding Goal-Post

(However, this rule is not the same in case of goal post as we know that the ball is considered 'in play' after bouncing off of a goal post, cross bar, corner flag, linesmen or referee if the ball remains on the playing field).

5.4   What is a 'Missed Goal-Shot' (MGS)?

The term 'goal-shot' in CFE has been defined in Sect. 5.1 above. By a 'Missed Goal-Shot' (MGS) in the Theory of CFE we mean that the goal-shot was an attempt for goal, but marginally missed the success and hence goes outside the 'Goal Post' getting into the net of the 'Goal-Area Post'. Needless to mention that the ball must cross the goal line completely. It is assumed that the goal-shot corresponding to a MGS is valid as per Laws of the Game, even not invalidated due to handballl, offside, etc. See Fig. 6 for a goal-shot which is a MGS.

If after a goal-shot the ball does not enter into the net of the 'Goal-Area Post' but touches/collides the 'Goal-Area Post' (either any vertical post(s) or horizontal crossbar or both) then this shot will also be called a MGS, even if the ball remains on the playing field after colliding the 'Goal-Area Post'.

However, if by a shot the ball touches (collides) the Goal Post first and then the Goal-Area Post, then the first collision will be considered as per CFE evaluation method as the valid and significant parameter (as mentioned in Sect. 7 here), ignoring the second collision which is insignificant. Consequently, such a shot does not qualify to be a kind of MGS. **Any goal shot, if qualified to be a SCB (as per definition of SCB given in Sect. 7.2) will <u>never</u> be qualified to be a MGS even if it enters into or does not enter into the net of Goal-Area Post.** One must be careful to note this point on MGS.

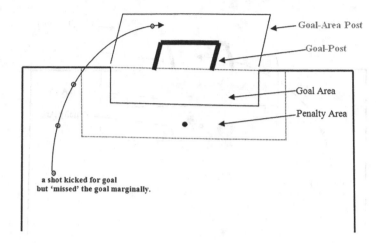

Fig. 6 Snapshot of a 'MGS'

5.5   *What is a 'Bad Goal-Shot' (BGS)?*

By a 'Bad Goal-Shot' (BGS) in the Theory of CFE we mean that the goal-shot was an attempt for goal, but it is neither a goal nor a MGS. Needless to mention that the ball must cross the goal line completely. It is assumed that the goal-shot corresponding to a BGS is valid as per Laws of the Game, even not invalidated due to handballl, offside, etc. See Fig. 7 for a goal-shot which is a BGS.

However, the shot in Fig. 8 is neither a MGS nor a BGS. It is a G.

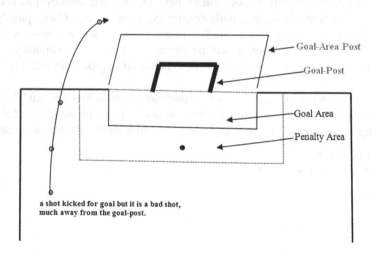

Fig. 7 Snapshot of a 'BGS'

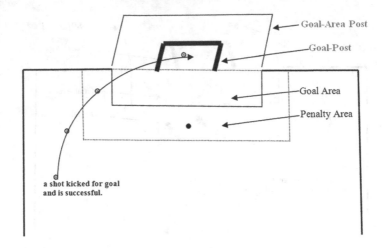

Fig. 8 A goal-shot which is a G

Any goal shot, if qualified to be a SCB (as per definition of SCB given in Sect. 7.2) will <u>never</u> be qualified to be a BGS even if it enters into or does not enter into the net of Goal-Area Post. One must be careful to note this point on BGS.

6 Categories of Fouls in CFE

Fouls committed during 90 min of football play are very common events. It is fact that some of the fouls are inadvertently committed, some are not. Consequently, the proper evaluation of such type of fouls can not be made without soft-computing approach. The 'Fuzzy Set Theory' will be the most suitable soft-computing tool for application in these cases which is the new powerful approach in our Theory of CFE.

Before introducing the notations for various categories of Fouls in our Theory of CFE, we quote below few important points from the 'Laws of the Game' [26–29]. According to the Law-12 of the Game, there are two kinds of fouls in soccer:

- Penal or Major Fouls.
- Non-Penal or Minor Fouls.

6.1   There are Nine Types of Penal or Major Fouls

These fouls are those which are committed intentionally and may result in a "Red Card". These fouls are as follows:

  (i)   Kicking a player.
 (ii)   Jumping up at a player.
(iii)   Charging a player in a rough way.
 (iv)   Charging a player from behind.
  (v)   Tripping a player.
 (vi)   Hitting or spitting at a player.
(vii)   Pushing a player.
(viii) Holding a player.
 (ix)   Handling the ball. (Except by a goalkeeper).
        This foul is called if the player is trying to control the ball with his hands or arms.

If one of these nine penalty fouls is committed and the referee blows his whistle and calls a foul, the opposing team gets a direct free kick. A "direct" kick means the opponent can try to score a goal directly from the kick. If the player committing the major foul receives a "red card" from the referee, he must leave the game, and is not allowed to return. No substitute player is allowed to play in such a case.

6.2   There are Five Types of Non-penal or Minor Fouls

If a player commits a minor foul he may receive a "Yellow Card" from the referee. The five minor fouls are as follows:

  (i)   Dangerous play.
        (Examples of a dangerous play are: high kicking near another player's head, or trying to play a ball held by a goalie).
 (ii)   Fair charging, but with the ball out of playing distance.
(iii)   Illegal obstruction. When a player intentionally takes a position between the ball and an opponent, when not within playing distance of the ball.
 (iv)   Charging the goalkeeper in the goal area.
  (v)   Goalkeeper Infringements fouls. Three types are:

        • Goalkeeper taking more than four steps while controlling the ball.
        • Goalkeeper playing the ball with his hands when the ball is kicked by a teammate.
        • Intentionally wasting time.

When the referee stops play by blowing his whistle for a minor foul, the opposite team is awarded an indirect free kick. A goal cannot be scored directly from an indirect free kick. The ball must be played by a player other than the one taking the indirect kick, before a legal goal can be scored.

6.3 Misconduct

There are two kinds of misconduct:

(i) When an action results in a caution or a "yellow card" from the referee. A referee may warn a player to improve his conduct before a caution is issued.

(ii) When an action results in a player being ejected from the game, a "red card". The referee has also the authority to "red card" coaches or spectators because of misconduct or interference of the game.

6.4 Fouls in the Theory of CFE

In our Theory of CFE, all the above FIFA rules (mentioned in Sects. 6.1, 6.2 and 6.3 above) are well applicable. However, in the Theory of CFE the Fouls in a football match are divided into three categories (Fig. 9):

(i) **Category F1**: Simple Foul (not shown any card by the Referee)
(ii) **Category F2**: Shown '**Yellow Card**' Foul
(iii) **Category F3**: Shown '**Red Card**' Foul

The 'simple fouls' may have various nature and amount of foul-gravity. Similarly the 'yellow-card fouls' may have various nature and amount of foul-gravity, and also the 'red-card fouls' may have various nature and amount of foul-gravity. These are in fact soft kind of parameters and an excellent Referee evaluates their grades by his best possible intellectual capability and judgment regarding the amount of foul gravity.

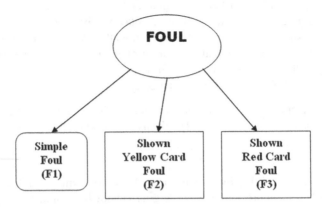

Fig. 9 Three categories of Fouls in a football match in CFE

As FIFA recruits the world's best Referees for the football matches, the decision of these Referees are regarded to be the best decisions and hence regarded to be the final decisions in every respect of their proceedings during play-time of any match. In a match, corresponding to a foul of any of the three categories F1 or F2 or F3, the foul-gravity is a very important but an imprecise and ill-defined term which can only be estimated by the Referee by his best possible intellectual judgment on the basis of finite number of significant criteria.

For instance, FIFA/UEFA could choose the following criteria to estimate the gravity of any foul (F1 or F2 or F3):

(i)   'bad intention, mainly for making tactful physical collision',
(ii) 'unfair way of ball possession',
(iii) 'inappropriate body language',
(iv) 'argument with the opponent player(s)', and
(v)   'argument with the Referee'.

where fouls are charged as mentioned in Sects. 6.1, 6.2 and 6.3 of FIFA rules.

However, the categories of fouls, the above list of criteria for estimating the gravity of any foul, and the list of parameters (which are introduced in Sect. 7 in this article) may be revised or improved time to time by FIFA(IFAB) experts and UEFA experts, and hence they are not absolutely fixed for all time in the Theory of CFE.

7   Parameters in the "Theory of CFE" for the 90 min Play Duration

FIFA World Cup is one of the most prestigious events happen on this earth! Almost all the countries in the world enjoy FIFA matches with heart-felt interest. But today, in the availability of modern science and advanced computational theories, we need to think about a basic question:

What are the parameters on which one can decide the true "Winner" of a football match?

Should it be just one parameter only? (which is the 'm-n goal' score at the end of 90 min of play which is a one time consolidated data on a single parameter only!).

Can we ignore the other significant and important parameters which show continuous performance, merits and demerits of each team during 90 min of play?

In our proposed Theory of CFE, there are 17 highly significant parameters as its components for continuous inputs to its software called by 'CFE-Software'. However, the parameters and the number of parameters are not absolutely fixed; these can be adjusted by the FIFA experts.

The parameters are of two kinds: Positive parameters and Negative parameters. The parameters in the Theory of CFE are defined as below:

7.1 Negative Parameters in CFE (Corresponding to Each Team)

There are 17 parameters (for each team) of real time nature considered in the theory of CFE for the 90 min play duration. Some of them are called **negative parameters** and rest of them are to be called **positive parameters**. Positive parameters are those which can extract the merits of a team and negative parameters are those which can extract the demerits of a team. There are seven negative parameters and ten positive parameters which are listed below.

However, more number of decision contributing parameters of real time nature, positive as well as negative, can be added in the theory of CFE in future if decided by FIFA (IFAB) and UEFA. The proposed list of parameters presented is not an absolutely fixed list for all time.

Negative Parameters (corresponding to each team) in CFE

There are seven negative parameters in football match for each team for the 90 min play duration. For a team, these are listed below:

(i) $F1$ = Number of 'Simple Fouls' committed (Not shown any card) by this team during the 90 min of play.

(Note: In the previous Sect. 6.4 we used the notation $F1$, $F2$ and $F3$ to denote the category titles of various kinds of fouls. Let us use the same notations here to denote their respective frequency too, if there is no confusion).

(ii) $F2$ = Number of 'Yellow Cards' shown to this team by the Referee during the 90 min of play.

(iii) $F3$ = Number of 'Red Cards' shown to this team by the Referee during 90 min of play.

(iv) O = Number of 'Offsides' committed by this team during the 90 min of play

(v) H = Number of 'Handballs' committed during the 90 min of play.

(vi) BGS = Number of 'Bad Goal Shots' performed by this team during the 90 min of play.

(It may be noted as mentioned earlier in Sect. 5.5 that a goal shot is called a BGS because it does not add any positive taste or positive interest to the football fans as well as to the football itself, whereas the MGS does add. Any BGS, which could not become a G or at least a MGS, is surely due to misappropriate timing of shot or due to shortfall in art/skill for that shot or due to some kind of lack in quality for that shot. And therefore the BGS is treated as a negative parameter.)

(vii) R = Number of Replacement of players made by the Coach (who are **not injured**) during 90 min of play for this team. Replacement of any injured players is **not counted** in this account of R.

7.2 Positive Parameters (Corresponding to Each Team) in CFE

There are ten positive parameters for each team for the 90 min play duration which are listed below.

(i) G = Number of Goals scored by this team during the 90 min of play.

(ii) BPC = 'Percentage of Ball Possession' by this team at the complete ground during the 90 min of play. Mathematically, the BPC is a continuous variable.

(iii) BPH = 'Percentage of Ball possession' by this team at the "half-ground of the opponent" during 90 min of play. Mathematically, the BPH is a continuous variable.

(iv) SCB = Number of Goal-Shots of this team without scoring Goal during 90 min of play but

   (i) which Collide at Goal Post (as defined in Sect. 5.2), or
   (ii) having a touch with the goalkeeper but without no touch by the opponent player in between, or
   (iii) having a touch both with the goalkeeper and Goal Post but without no touch by the opponent player in between.

(v) MGS = Number of 'Missed Goal Shots' by this team during the 90 min of play.

(vi) CK = Number of Corner-kicks (CK) availed by this team during 90 min play. It may be noted that a goal-shot may be qualified to be categorized in both SCB and CK, and in that case both are to be considered.

(vii) T = Number of Throws availed by this team during 90 min of play.

(viii) 2G = Consecutive two goals scored by this team (without any goal scored by the opponent in-between these two goals) during 90 min of play.

(ix) 3G = Consecutive three goals scored by this team (without any goal scored by the opponent in-between these three goals) during 90 min of play.

(x) nG = Consecutive four or more goals (n > 3) scored by this team (without any goal scored by the opponent in-between these n number of goals) during 90 min of play.

8 CS Value of a Foul (F1 or F2 or F3)

The **CS** abbreviation stands for CFE-Score. For each team in a game, corresponding to every parameter there is a 'CS value' which goes to finalize the 'CS score' of the team (explained in Sect. 12). The CFE method is based on computation of several CS values. Let us recollect that in a fuzzy set A of a universe of discourse U, the decision maker awards a grade of membership to each element from the interval [0,1] by his best possible intellectual judgment and knowledge.

Fouls are negative parameters in the Theory of CFE as mentioned in Sect. 6.4 above. The CS value of a foul (of category F1 or F2 or F3) is always a crisp value. But for this, the referee initially awards a 'punishment fuzzy set' to this foul at real time of play by his best possible intellectual judgment about the gravity of the foul. The 'punishment fuzzy set' is awarded by the Referee using the 'fuzzy pocket machine' M (see Sect. 9). This fuzzy set is then de-fuzzified by one of the three algorithms: Algo-1, Algo-2, Algo-3, whichever be applicable depending upon the actual category of foul (F1 or F2 or F3) to get the crisp CS value of that foul. Detail explanation is given here in the subsequent subsections.

For a foul (which could be of one of the three categories: F1 or F2 or F3), the 'universal set' is the crisp set F given by $F = \{f_1, f_2, f_3, f_4, f_5\}$, where

f_1 = 'bad intention mainly for tactful physical collision',
f_2 = 'unfair way of ball possession',
f_3 = 'inappropriate body language',
f_4 = 'argument with the opponent player(s)', and
f_5 = 'argument with the Referee'.

Note: However, the elements f_i of the universal set F may be modified, number of elements may be increased/decreased and fixed by the FIFA experts or UEFA experts. For the sake of presentation here, we consider the five-membered hypothetical universal set F as mentioned above for the Fouls.

In the Theory of CFE, for every foul (F1 or F2 or F3) committed by a player (i.e. by a team), the Referee takes a fuzzy action against the concerned player (i.e. against the concerned team) by awarding a "punishment fuzzy set" A of the universal set F to the team. The 'punishment fuzzy set' A will be awarded to the team using the fuzzy pocket machine at real instant of time, prior to the foul-kick to be kicked by the opponent team. This will hardly take ten to twelve seconds time of the Referee, which is a quite reasonable amount of time to award a 'punishment fuzzy set'.

9   "Fuzzy Pocket Machine" M for the Referee

There is a handy small "fuzzy pocket machine" M for the Referee (see Fig. 10) using which the referee awards a punishment fuzzy set A of the universe F. The Referees need not be experts in fuzzy set theory; they can be easily trained within just 30 min of demonstration on: how to use the 'fuzzy pocket machine' to award a punishment fuzzy set. Almost similar type of "fuzzy pocket machine" we earlier introduced in [9, 12].

A fuzzy pocket machine M is a simple electronic wireless machine looking like a mobile phone. It is having its own memory. Once the Referee awards a 'punishment

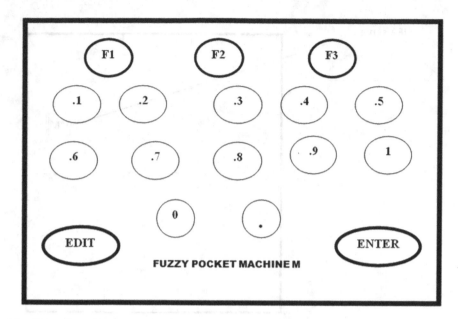

Fig. 10 A 'Fuzzy Pocket Machine' **M** for the Referee

fuzzy set', the data will be automatically stored immediately in the concerned database in the FIFA-Server for a software called by "CFE-Software", and a back-up is also automatically stored in the memory of the fuzzy pocket machine M itself.

The machine M will have the seventeen buttons (see Fig. 10) which are mentioned below:

(i)   three buttons in the name of F1, F2 and F3 respectively,

(ii) twelve buttons in the name of: . (dot), 0, .1, .2, .3, .4, .5, .6, .7, .8, .9, and 1 respectively, and

(iii) two more buttons in the name of 'ENTER' and 'EDIT'.

All these seventeen buttons of a fuzzy pocket machine M are shown in Fig. 10, and they are press buttons.

9.1   How to Input Data by the Referee Using His Machine M ?

The Referee inputs punishment fuzzy sets (see also Sect. 9.3) using his fuzzy pocket machine M by his best intellectual capability and knowledge which automatically get stored in the database at server (see Fig. 11) and a back-up is also automatically stored in the memory of the fuzzy pocket machine M itself.

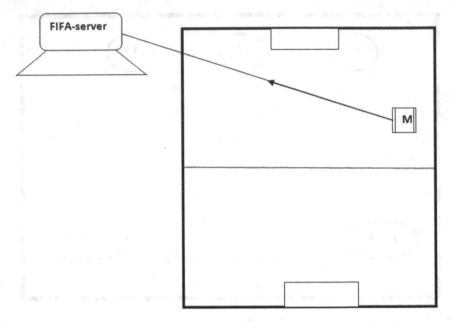

Fig. 11 Referee sends input to the database at the FIFA-server from his 'Fuzzy Pocket Machine' **M**

It is known that in any fuzzy set the grade or membership value of every object is a non-negative real number in the closed interval [0,1] proposed by the concerned decision maker (here the decision maker is the Referee).

The following is the procedure for giving input of a real number from the closed interval [0,1] using the machine M to the database at the server.

(i)   To input a data like 0.6, the Referee has to press .6 button only. To input the data 0, the Referee has to press the 0 button only. To input the data 1, the Referee has to press the 1 button only.

(ii)   However, to input a data like 0.63, the Referee has to press the following sequence of buttons: . (dot), .6, .3.

(iii)   Similarly, to input a data like 0.638, the Referee has to press the following sequence of buttons: . (dot), .6, .3, .8.

9.2   Key Architecture of the 'Fuzzy Pocket Machine' M

The architecture of the 'fuzzy pocket machine' M is so designed that on pressing the . (dot) button initially, the each of the other buttons .1, .2, .3, .4, .5, .6, .7, .8, .9 behaves 'without decimal point' till the 'ENTER' button be not pressed.

However, the machine M resumes its original status after the 'ENTER' button be pressed.

Besides that, a Referee can not input any membership value data which is greater than 1 or less than 0 or which has more than three decimal places (fuzzy pocket machine itself does not permit it by its own architecture and configuration).

The 'fuzzy pocket machine' M has its own memory which keeps back-up of all the real time data communicated to the FIFA-server (or UEFA-server) time to time by the Referee during the 90 min of play (see Fig. 10).

9.3 How to Award a "Punishment Fuzzy Set" Using the 'Fuzzy Pocket Machine'

The Theory of CFE being a continuous evaluation method, every amount of credit/ discredit of a player cater to the cumulative amount of credit/discredit of the concerned team as a whole. For a committed Foul during play, depending upon the gravity of Foul, the Referee follows the following sequence of steps to award a "punishment fuzzy set" A to the concerned player (i.e. to the concerned team) by his best possible intellectual judgment:

Step-1: Referee press one of the buttons F1 or F2 or F3 whichever be appropriate by his judgement, and then press the button 'ENTER'. (This signifies which of the three categories of fouls is awarded by the Referee).

Step-2: Referee inputs one membership value from the closed interval [0,1] using one or more buttons: . (dot), 0, .1, .2, .3, .4, .5, .6, .7, .8, .9, and 1, and then the button 'ENTER'.
(This is the membership value of the element f_1 for the fuzzy set A of the universal set F by best possible intellectual judgment of the Referee). In this regard the Sect. 9 may be revisited.

Step-3: Referee inputs one membership value from the closed interval [0,1] using one or more buttons: . (dot), 0, .1, .2, 0.3, .4, .5, .6, .7, .8, .9, and 1, and then the button 'ENTER'.
(This is the membership value of the element f_2 for the fuzzy set A of the universal set F by best possible intellectual judgment of the Referee).

Step-4: Referee inputs one membership value from the closed interval [0,1] using one or more buttons: . (dot), 0, .1, .2, .3, .4, .5, .6, .7, .8, .9, and 1, and then the button 'ENTER'.
(This is the membership value of the element f_3 for the fuzzy set A of the universal set F by best possible intellectual judgment of the Referee).

Step-5: Referee inputs one membership value from the closed interval [0,1] using one or more buttons: . (dot), 0, .1, .2, .3, .4, .5, .6, .7, .8, .9, and 1, and then the button 'ENTER'.

(This is the membership value of the element f_4 for the fuzzy set A of the universal set F by best possible intellectual judgment of the Referee).

Step-6:   Referee inputs one membership value from the closed interval [0,1] using one or more buttons: . (dot), 0, .1, .2, .3, .4, .5, .6, .7, .8, .9, and 1, and then the button 'ENTER'.

(This is the membership value of the element f_5 for the fuzzy set A of the universal set F by best possible intellectual judgment of the Referee).

While the ENTER button for Step-1 is pressed, the machine M itself is intelligent to keep count whether the ENTER button is pressed for five times more corresponding to the next five steps, and in the following sequence:

Grade value, ENTER;   Grade value, ENTER;   Grade value, ENTER;   Grade value, ENTER;   Grade value, ENTER.

In case the count value does not match within a reasonable time, the reminder message comes to the Referee. But it is expected that Referee will not forget to complete his input process of a punishment fuzzy set.

However, the Referee can use the EDIT button to change his recently input membership value corresponding to a 'punishment fuzzy set' for which the ENTER button is not yet pressed by him, although such cases will be rare by the talent Referees. But after editing (if done), he has to press the ENTER button in order to save the modification in the memory of the FIFA-server. These inputs of the Referee are transmitted to the database in the server directly from the playground at real instant of time automatically. A Referee can award a punishment fuzzy set in just 10–15 seconds of time.

10   Three Algorithms Algo-1, Algo-2 and Algo-3 for De-fuzzification

Corresponding to the three categories of fouls (F1, F2, F3), there are three respective cases and algorithms which are explained below. These algorithms (of CFE-software) de-fuzzify the fuzzy inputs given by the Referee.

Case-1:   If the Foul is a Simple Foul

This case is for a foul of category F1 which has been decided and input by the Referee at some instant of time. The CS value of a foul F1 denoted by CS(F1) will be always one of the three integers 2, 3 and 4.

Let A be the 'punishment fuzzy set' awarded by the Referee corresponding to this foul committed by a team. Then the following algorithm called by Algo-1 will be applicable on the basis of α-cut of the fuzzy set A. Thus the Algo-1 is applicable if the foul is of category F1 only, and for $\alpha = 0.5$.

Algo-1

If the 0.5-cut of A (i.e. the set $A_{0.5}$) is a null set
        then CS(F1) = 2, Stop. *else*
If the 0.8-cut of A (i.e. the set $A_{0.8}$) is a null set
        then CS(F1) = 3, Stop. *else*
CS(F1) = 4.

Case-2:   If the Foul is a Yellow Card Foul

This case is for a foul of the category F2 which has been decided and input by the Referee at some instant of time. The CS value of a foul F2 denoted by CS(F2) will be always one of the three integers 4, 5 and 6.

Let A be the 'punishment fuzzy set' awarded by the Referee corresponding to this foul committed by a team. Then the following algorithm called by Algo-2 will be applicable on the basis of α-cut of the fuzzy set A. Thus the Algo-2 is applicable if the foul is of category F2 only, and for $\alpha = 0.5$.

Algo-2

If the 0.5-cut of A (i.e. the set $A_{0.5}$) is a null set
                then CS(F2) = 4, Stop. *else*
If the 0.8-cut of A (i.e. the set $A_{0.8}$) is a null set
                then CS(F2) = 5, Stop. *else*
CS(F2) = 6.

Case-3:   If the Foul is a Red Card Foul

This case is for a foul of the category F3 which has been decided and input by the Referee at some instant of time. The CS value of a foul F3 denoted by CS(F3) will be always one of the three integers 6, 7 and 8.

Let A be the 'punishment fuzzy set' awarded by the Referee corresponding to this foul committed by a team. Then the following algorithm called by Algo-3 will be applicable on the basis of α-cut of the fuzzy set A. Thus the Algo-3 is applicable if the foul is of category F3 only, and for $\alpha = 0.5$.

Algo-3

If the 0.5-cut of A (i.e. the set $A_{0.5}$) is a null set
                    then CS(F3) = 6, Stop. *else*
If the 0.8-cut of A (i.e. the set $A_{0.8}$) is a null set
                    then CS(F3) = 7, Stop. *else*
CS(F3) = 8.

Note: There should not be any confusion on the fact that the CS value 4 is common to Case-1 and Case-2, and that the CS value 6 is common to Case-2 and Case-3. It may be recollected from the Fuzzy Set Theory that the graphs of the membership functions of 'High', 'Medium' and 'Low' are not non-intercepting or disjoint graphs; and here too we have the analogous philosophy by which the CS value 4 is common to Case-1 and Case-2, and that the CS value 6 is common to Case-2 and Case-3. It is one of the most beautiful notions Prof. Zadeh introduced in his Fuzzy Set Theory.

11   CS Values of Other Parameters in CFE During the 90 min Play

It is already mentioned that the duration of play in CFE is absolutely fixed which is 90 min, not more or not less in almost all games (except CPS Play which is having a very very rare possibility, almost nil).

As mentioned earlier that in CFE method, there is **no** 'extra time of play' beyond 90 min, there is **no** 'Replay' beyond 90 min, there is **no** 'Penalty Shootout', there is **no** 'CPS' in almost all the CFE cases (very low chance), and there is **no** 'sudden death' round. This is the revolutionary merit and huge dominance of the CFE method over the existing method of FIFA/UEFA.

Out of seventeen parameters in the theory of CFE for the 90 min play duration, there are seven number of negative parameters and ten number of positive parameters. The CS values of these 17 parameters play a vital role in the Theory of CFE. The CS values of three negative parameters F1, F2 and F3 are discussed in the preceding subsection. In this section we discuss the CS values of all other parameters.

11.1   CS Value of Negative Parameters (of a Team) in CFE

It is discussed in Sect. 10 earlier that during the 90 min of play, for a Simple Foul F1, the CS(F1) $\in$ {2,3,4}. For a Yellow Card Foul F2, the CS(F2) $\in$ {4,5,6}. And for a Red Card Foul F3, the CS(F3) $\in$ {6,7,8}.

For the other negative parameters, we propose to fix the following CS values in the Theory of CFE. **However, these CS values may be reset by the FIFA experts or UEFA experts.**

- (i)   For each Offside, CS(O) = 1;
- (ii) For each Handball, CS(H) = 1;
- (iii) For every Bad Goal Shot, CS(BGS) = 1;
- (iv) For Replacement (R) of players who are **not injured** during play, the CS value is given by the following norm:

$$CS(R) = 1, \text{ if 1 or 2 players are Replaced.}$$
$$ = 2, \text{ if 3 or more players Replaced.}$$

(It is to be noted that the number of replacements of injured players is not considered in the Theory of CFE for computing any CS value).

11.2 CS Value of Positive Parameters (of a Team) in CFE

For the positive parameters of a team for the 90 min play duration, we propose to fix the following CS values in the Theory of CFE. **However, these CS values may be reset by the FIFA experts or UEFA experts.**

- (i)   For each Goal G scored, CS value CS(G) = 10.
- (ii) If the Ball Possession of a team across the complete ground the during 90 min of play is x%, then CS(BPC) = x/20.
- (iii) If the Ball Possession of a team across the "Half-ground of the Opponent" during 90 min of play is y%, then CS(BPH) = y/10.
- (iv) For each Shot which Collides at Goal Post or having a touch with the goalkeeper or both, without scoring Goal during 90 min of play, the CS value is CS(SCB) = 1.5.
        (However, if a Shot Collides at Goal Post with a Goal, then there is no CS (SCB) value for this shot, because in such a case SCB is irrelevant).
- (v)   For every Missed Goal Shot, CS(MGS) = 1;
- (vi) For each corner kick (CK) during 90 min of play, the CS value is given as below:
        CS(CK) = 1, if the ball of that corner-kick goes directly outside the play ground without touching any player or any location inside the playground or goal post.
        CS(CK) = 2, otherwise, for all other cases.
        It is to be noted that in case the corner kick be a SCB then the CS(SCB) value will also be considered as usual, in addition to due value of CS(CK). However, if a corner kick be a Goal, then there is no CS(SCB) value for this shot, but the due value of CS(CK) will not be ignored and hence will be credited.
- (vii) If a team gets a Throw, then for each such throw the CS value is given by CS(Throw) i.e. CS(T) = 1

(viii) If a team scores two consecutive goals (without any goal scored by the opponent in-between these two goals) during 90 min of play, then for each such case the CS value is given by CS(2G) = 1, in addition to the normal CS(G) value for each goal.

(ix) If a team scores three consecutive goals (without any goal scored by the opponent in-between these three goals) during 90 min of play, then for each such case the CS value is given by CS(3G) = 5, in addition to the normal CS(G) value for each goal. In such case the CS value of 2G will not be considered.

(x) If a team scores n (n > 3) consecutive goals (without any goal scored by the opponent in-between these n goals) during 90 min of play, then for each such case the CS value is given by CS(3G) = 10, in addition to the normal CS(G) value for each goal. In such case the CS value of 2G or 3G will not be considered.

12 'CS Score' of a Team at Any Time During the 90 min Play-Time

In the earlier sections we defined CS values of all the positive and negative parameters of a team during the 90 min play-time. Suppose that there is a FIFA/UEFA football game between the two teams X and Y. Then the CFE-software can be executed in the FIFA-Server to compute the individual 'CS scores' of both the teams X and Y at any real instant of time during 90 min of play.

To compute the CS Score of a Team (say, team X) at any instant of time, the CFE-software at the FIFA-server computes the following values first of all:

(i) Total amount of CS values accumulated so far corresponding to all the negative parameters for the team X which is denoted by **NCS(X)**.

(ii) Total amount of CS values accumulated so far corresponding to all the positive parameters for the team X which is denoted by **PCS(X)**.

The "CS Score" of the team X at any real instant of time (during the 90 min play-time) is computed at the FIFA-server using the simple mathematical formula:

$$CS(X) = 1000 + PCS(X) - NCS(X).$$

Note: In the above formula for CS(X), there is no significance of the amount 1000. It is added for no other purpose but just to ensure that the value of CS(X) does not become a negative real number under any circumstances. If for a team X it happens at the end that CS(X) < 1000, then it obviously signifies that the team X is a poor team in terms of football skills and merits.

The CS-score CS(X) of a team is a function of several parameters, some of them are positive parameters and some of them are negative parameters. **It is to be noted** that, **mathematically the CS-score of each team varies continuously at every second because of the fact that some of the parameters (ex. BPC, BPH, etc.) of this function are <u>continuous</u> variables.** However, as mentioned in this article, that the Electronic Display Boards will update for both the teams the latest CS scores nine times during 90 min of play, once in every 10 min, by executing the CFE-software nine times only (if there is no CPS play).

It may be noted that the term "CS value" is used for evaluation of a parameter whereas the term "CS score" is used for evaluation of a team as a whole. But both are basically CS i.e. CFE-Score. Besides that, in the term "CS score", the word 'score' appears twice if expanded; but let us ignore this grammatical error for the sake of smooth practice of the analysis while in reality.

Finally at the end of 90 min play the CFE-software computes the "Winner" of this football game as mentioned in the next section.

13 Who is the "Winner" by CFE Method?

Suppose that the 90 min play is just over with m-n goals score, and the last whistle is blown by the Referee. Then the CFE-software is executed in the FIFA-Server to compute the individual 'CS scores' of both the teams X and Y. The question now is: Who is the "Winner" in today's match? The following is the method for making the declaration according to the Theory of CFE.

Declaration

If $CS(X) > CS(Y)$ then the team X is the 'better' team, and hence the team X is declared to be the **Winner** by FIFA.

Otherwise, if $CS(X) < CS(Y)$ then the team Y is the 'better' team, and hence the team Y is declared to be the **Winner** by FIFA.

In CFE method, there are 17 parameters for the 90 min play duration. Positive parameters are those whose quantified values add to the credit of a team and negative parameters are those whose quantified values add to the discredit of a team. Consequently a poor performer team can not escape from the CFE evaluation method to claim for 'Winner' even if the team can manage, say 1-0 goal ahead, at the end of 90 min of play. This is a breakthrough philosophy incorporated in the CFE evaluation method, unlike the existing method of FIFA/UEFA.

14 A Very Very Rare Case "TCFE": Whose Possibility is not just Zero(0) Mathematically

A <u>very very rare case</u> is that at the end of 90 min play of the game between the teams X and Y, if by any chance it happens that

$$CS(X) = CS(Y).$$

This is called to be the case of "Tie CFE" or **TCFE**.

Mathematically the possibility of an occurrence of such a **TCFE** case is not absolutely nil. But it is quite obvious that such a situation is so much low probable that it may not even happen once in 1000 played games!. However, in that case any one of the following three methods TCFE-1, TCFE-2 and TCFE-3 is to be applied (as be decided by the concerned organizing committee in the set of Rules):

TCFE-1: 20 min extra play (10 min + 10 min) and with application of CFE during this 20 min of play.
Or
30 min extra play (15 min + 15 min) and with application of CFE during this 30 min of play.

TCFE-2: CPS play. The CPS play is explained later in this article in Sect. 19.

TCFE-3: It is a sequential combination of TCFE-1 and TCFE-2. In other words, TCFE-3   means first TCFE-1 play and then TCFE-2 play will take place.

Althrough in our present article, we will prefer TCFE-2 only (neither TCFE-1 nor TCFE-3), in case of TCFE play in CFE.

14.1   Who is the "Winner" by TCFE in the CFE Method (if Situation Arises)?

If **TCFE** play is required by any chance, all the CS values of positive and negative parameters for both the teams generated during the 90 min play are ignored and the CS scores of both the teams of 90 min play are also ignored. The decision is taken from the inputs out of TCFE play time only. Although we prefer **TCFE-2** as the ideal method, nevertheless we explain below all the three methods:-

If **TCFE-1** is applied, then the final decision is to be taken on the basis of the revised CS scores of both the teams X and Y. The team having higher CS score is the 'better' team and hence declared to be the "Winner".

Mathematically, still there exists possibility of arriving at equal CS scores, and in that case any one of the three methods TCFE-1, TCFE-2 and TCFE-3 is to be applied (as be decided by the concerned organizing committee) and so on, until the CFE-software can compute the 'better' team.

If **TCFE-2** is applied, then the final decision is to be taken on the basis of the CS scores of CPS play only, for both the teams X and Y. The details of TCFE-2 (CPS play) is explained later in Sect. 19. The team having higher CS score is the 'better' team and hence declared to be the "Winner".

Mathematically, still there exists possibility of arriving at equal CS scores, and in that case TCFE-2 is to be repeated once more, and so on, until the CFE-software can compute the 'better' team.

If **TCFE-3** is applied, then the final decision is to be taken on the basis of the '**Cumulative CS Scores**', where the 'Cumulative CS score' of a team is calculated by adding the CS-scores of that team earned from TCFE-1 play and from TCFE-2 play. The team having higher 'Cumulative CS Score' is the 'better' team and hence declared to be the "Winner".

Mathematically, still there exists possibility of arriving at equal CS scores, and in that case only TCFE-2 is to be repeated once more, and so on, until the CFE-software can compute the 'better' team.

15   A Hypothetical Example of a Football Game of FIFA

We explain here with <u>hypothetical</u> data an application of this fuzzy CFE method in a football game played between two teams X and Y in a FIFA match (imaginary match) which has ended with 4-3 goals after 90 min of play. The existing method of FIFA/UEFA chooses the 'Winner' team <u>directly</u> on the basis of one and only one parameter which is the '4-3 goal score' at the end of 90 min of play, ignoring many other important and significant parameters of the 90 min play duration which continuously exist well since the start of the game. Our proposed method CFE does not focus at declaring the 'Winner' directly, but at finding the actually 'BETTER' team by <u>continuous</u> evaluation and by computation of all the data/information available from the field since the start of the game for 90 min, but one out of many of those important data being the '4-3 goal' score.

According to the Theory of CFE, the data from the field go to the database in FIFA server as the input components for the CFE-software in this continuous evaluation method. Suppose that the following is the statistics (R-Statistics [11]) for the two teams X and Y as recorded in the database of the FIFA-server corresponding to the 17 parameters of CFE method, as shown in Table 1 which are the final data at the end of 90 min of play. The database in FIFA server does also contain all the punishment fuzzy sets awarded by the Referee corresponding to the fouls recorded in today's match.

15.1   Fouls Committed by the Team X

The Table 1 shows that in this FIFA match in total there are five fouls committed by the team X, out of which there are three F1 fouls, two F2 fouls and nil number of F3 fouls.

The punishment fuzzy sets corresponding to the three F1 fouls committed by the team X are awarded by the Referee via his fuzzy pocket machine M which are as below:

Out of three number of F1 fouls committed, the punishment fuzzy set awarded corresponding to the first F1 is:

Table 1 The statistics of 17 parameters for team X and team Y at the end of 90 min play

Serial No.	Parameters	Frequency (X): Number of occurrences for team X	Frequency (Y): Number of occurrences for team Y
1	F1	3	1
2	F2	2	1
3	F3	0	0
4	O	3	4
5	H	3	2
6	BGS	1	0
7	R	3	1
8	G	4	3
9	BPC	40%	60%
10	BPH	26%	50%
11	SCB	0	0
12	MGS	7	6
13	CK	3	5
14	T	6	9
15	2G	0	0
16	3G	0	0
17	nG (n > 3)	0	0

X(i):   the fuzzy set $\{(f_1, 0.85), (f_2, 0.8), (f_3, 0.95), (f_4, 0.85), (f_5, 0.9)\}$, and

The punishment fuzzy set awarded corresponding to the second F1 is:

X(ii):   the fuzzy set $\{(f_1, 0.95), (f_2, 0.85), (f_3, 0.9), (f_4, 0.95), (f_5, 1)\}$, and

The punishment fuzzy set awarded corresponding to the last F1 is:

X(iii):   the fuzzy set $\{(f_1, 0.85), (f_2, 0.6), (f_3, 0.95), (f_4, 0.85), (f_5, 0.8)\}$.

The punishment fuzzy sets corresponding to the two F2 fouls committed by the team X are given by the Referee via his fuzzy pocket machine M which are as below:

X(iv):   the fuzzy set $\{(f_1, 0.95), (f_2, 0.9), (f_3, 0.95), (f_4, 1), (f_5, 0.95)\}$, and
X(v):   the fuzzy set $\{(f_1, 0.85), (f_2, 0.8), (f_3, 0.95), (f_4, 0.9), (f_5, 1)\}$.

While CFE-software executes its Algo-1 and Algo-2, it will get the following de-fuzzified CS values for the team X:

$CS(X(i)) = 4$, $CS(X(ii)) = 4$, $CS(X(iii)) = 4$, $CS(X(iv)) = 6$, $CS(X(v)) = 6$.

There is no F3 fouls committed by any player of team X and hence the Algo-3 of CFE-software will not be executed for the team X for this match.

15.2   Fouls Committed by the Team Y

The Table 1 also shows that in this FIFA match there are two fouls committed by the team Y in total, out of which there is one F1 foul, one F2 foul and nil number of F3 fouls.

The punishment fuzzy set corresponding to the one F1 foul committed by the team Y is given by the Referee via his fuzzy pocket machine M which is as below:

Y(i):   the fuzzy set $\{(f_1, 0.2), (f_2, 0.1), (f_3, 0.15), (f_4, 0), (f_5, 0)\}$.

The punishment fuzzy set corresponding to the one F2 foul committed by the team Y is given by the Referee via his fuzzy pocket machine M which is as below:

Y(ii):   the fuzzy set $\{(f_1, 0.15), (f_2, 0.2), (f_3, 0.1), (f_4, 0), (f_5, 0)\}$.

While CFE-software executes its Algo-1 and Algo-2, it will get the following de-fuzzified CS values for the team Y:

$CS(Y(i)) = 2$ and $CS(Y(ii)) = 2$.

There is no F3 fouls committed by any player of team Y and hence the Algo-3 of CFE-software will not be executed for team Y for this match.

15.3   The CS Values of 17 Parameters for Team X and Team Y

The CFE-software computes the CS values of all the 17 parameters for both the teams X and Y nine times, once in every ten minutes, during the 90 min of play.

The hypothetical example below shows the last execution of CFE-software to compute the CS values of all the 17 parameters for both the teams X and Y at the end of 90 min of play, which are shown in the Table 2 (as per specification mentioned in Sect. 11 earlier).

The CFE-software then computes the **NCS** and **PCS** values of each team as below:

$NCS(X) = 12 + 12 + 0 + 3 + 3 + 1 + 2 = 33$ and
$PCS(X) = 40 + 2 + 2.6 + 0 + 7 + 5 + 6 + 0 + 0 + 0 = 62.6$.

$NCS(Y) = 2 + 4 + 0 + 4 + 1 + 0 + 1 = 12$ and
$PCS(Y) = 30 + 3 + 5 + 0 + 6 + 10 + 9 + 0 + 0 + 0 = 63$.

The CFE-software then computes the 'CS Score' of each team as below:

$CS(X) = 1000 + PCS(X) - NCS(X) = \mathbf{1029.6}$ and
$CS(Y) = 1000 + PCS(Y) - NCS(Y) = \mathbf{1051}$.

Table 2 The CS values of 17 parameters for team X and Y at the end of 90 min play

Serial No.	Parameters	Frequency (X)	CS value (for X)	Frequency (Y)	CS value (for Y)
1	F1	3	12	1	2
2	F2	2	12	1	4
3	F3	0	0	0	0
4	O	3	3	4	4
5	H	3	3	1	1
6	BGS	1	1	0	0
7	R	3	2	1	1
8	**G**	**4**	**40**	**3**	**30**
9	BPC	40%	2	60%	3
10	BPH	26%	2.6	50%	5
11	SCB	0	0	0	0
12	MGS	7	7	6	6
13	CK	3	5	5	10
14	T	6	6	9	9
15	2G	0	0	0	0
16	3G	0	0	0	0
17	nG (n > 3)	0	0	0	0

15.4   Final Output of the CFE-Software

Since CS(Y) > CS(X) in this match, the CFE-software computes that the team Y is the 'better' team and hence declared to be the **Winner**, the team X is the looser.

Declaration: The team Y is the '**Winner**' in today's FIFA football match.

Note: It is to be noted that the goal results in this example is 4-3 at the end of 90 min play, i.e. the team X scored 4 goals whereas the team Y scored 3 goals. By FIFA rule the 'Winner' in this game is X whereas by our CFE rule the 'Winner' in this game is Y. It is because of the reason that FIFA rules say X is the better team in today's match whereas the CPS rules say Y is the better team. Any football fan or football analyst will surely agree with the decision computed by the CFE-software, not with the decision of FIFA.

16   Comparing CFE with the Obsolete FIFA/UEFA Rules with Respect to the above Example

In the above example of football game of 'Team-X versus Team-Y', see that **IF** the fuzzy CFE method is not applied to this football match played between the two good teams X and Y, then by the existing FIFA/UEFA norms the FIFA will declare the team X as the **Winner (by 4-3 goals)** and the team Y as the **looser**.

But any football-fan or expert witnessing this FIFA match will get psychologically shocked with this declaration of Winner by FIFA, witnessing the 90 min continuous play of both the teams at the stadium and also visualizing the continuous performance of each and every player of both the teams.

This poor decision (of declaring the team X as the winner) is not any fault of FIFA (IFAB) or UEFA nor of any of the teams X and Y, nor of the Referee, nor of any football fans, nor of the venue, nor of the host country, nor of the weather-climate or any other reasons.

Surely it is due to an <u>obsolete method</u> presently being followed by FIFA (IFAB) and UEFA on one of the most important issues of football game: "How to select the 'Winner' if the result comes with 'm-n goals' after 90 min of play, where either m = n or m $\neq$ n?".

It is due to <u>non-availability</u> of any new innovative modern mathematical (and philosophical) technique which can provide by <u>continuous</u> evaluation of the match performance a truly correct solution and result, retaining the justice to the football, retaining the interest of the game, retaining the interest of the football fans and experts, retaining the interest of the football world as a whole. The author here genuinely claims that the soft computing method CFE is a much improved method for the football subject, can provide a huge amount of justice and fairness to the football game.

17   Optimized Benefits in "CFE"

In the "**Theory CFE**" there are many in-built <u>optimization</u> of football elements. The following benefits can be easily achieved by default, without any additional cost of anything. These are not possible in the existing rules of FIFA(IFAB) or UEFA:

(i)   The proposed method "CFE" is a soft-computing method based upon 'continuous evaluation of 90 min'. The proposed "CFE" is an <u>integrated</u> approach for evaluation and judgment unlike the existing approach of evaluation which is of 'one time' nature. The 'one time' nature of the existing deciding method of FIFA(IFAB) and UEFA is the main source of weakness. In our continuous evaluation method, enhancement in the 'quality of discipline' by the players during play can be <u>optimized</u> by a huge amount compared to the present situation. It is because of the reason that every player knows that his every second's performance, every second's discipline, etc. cater quantitatively to the final decision of the game. Consequently, every player will surely remain more stick to the football game by his highest amount of discipline and honesty, not to any odd intention.

(ii) There will **not** be any necessity of 'extra time of play' after tie case of 'n-n goals' if happens after 90 min of play. The existing rule for 'extra time' can be totally discontinued in future. The existing rule for 'Penalty Shoot' can also be totally discontinued in future. The rule of the CPS in CFE is too rare case and hardly to happen even once in 1000 games!. These beautiful features will certainly provide a huge amount of optimization on satisfaction to the football fans, even to all the workers/staff/agents involved in the match.

(iii) The spirit of 'team sport' (football should not be individual sport) in football matches will be further optimized and justified, as CFE computes the data of the play by all thé 11 players of each team, even during CPS play if played.

(iv) The **most important** and the **extra-ordinary merit** of the "CFE" method is that both the Winner team and the Looser team, both the football fans and the organizing committee will be certainly and fully satisfied and convinced with the final judgment, without any element of doubt in mind. This **satisfaction** is 100% <u>guaranteed</u> by "CFE" method in each and every football match to all the following seven categories of people/structures:

(a) Winner team and their supporters
(b) Looser team and their supporters
(c) all football fans and analysts
(d) all the Referees
(e) Organizing Committee and other committees
(f) all the workers/staff/agents involved in the match
(g) the subject 'Football Science'.

Providing a fair judgment, fair decision, satisfaction, football spirit, and also a micro-level analysis to both the Winning team and loosing team in each and every FIFA/UEFA game is certainly possible if played using our proposed "CFE Method". The "CFE Method" will surely become a major breakthrough in the research area of 'Football Science', providing to FIFA and UEFA a genuine scope for huge improvement of the existing rules/laws and practices followed in the football games. The method is so constructed that it can be flexibly adjusted as many times as required in future with the support of new research findings on science and technology time to time.

18   Example of Few Top Class Important Games Which Could Have Certainly Been Given 'Better Decisions' by CFE Method

One of the very rich and strong merits of our proposed CFE method is that it <u>does not need</u> 'penalty shootout' for deciding the Winner of a football game that has even ended with 'n-n goals' after 90 min of play, and does not even need any 'extra play' or 'sudden death' play. Unfortunately, there are a large number of games played so far which had been poorly decided by FIFA/UEFA just because of non-availability of any other way-out for giving better decisions, appropriate decision or fair decision, except giving the decision by only one piece of last moment data which is the 'm-n goal' data. Such type of instances are very large in number in football history. Some of the recent instances are produced below for realization of the gravity of such unfortunate situations.

18.1 Cases from "UEFA Euro 2016 Games": Which Surely Missed an Appropriate Evaluation Method for Giving Better 'Final Decisions'

Extra Time play and/or Penalty shoot-out are very frequent in FIFA World Cup or UEFA matches. It is shocking fact that the inferior teams are tempted sometimes to play preferably for a scoreless draw or for any 'n-n goals' draw at the end of 90 min of play, and then even during extra time play, calculating that a shoot-out may offer their best hope of victory. Killing play-time as maximum as possible is not an impossible job or tough job of a team during 90 min of play, without violating FIFA rules of the game. In this section we furnish few UNFORTRUNATE examples (extracted from the open-source websites [26–29]) out of many.

Consider the recent **UEFA Euro-2016 games** held in France from 10th June to 10th July 2016. Let us focus at the performance of the champion team Portuguese in Euro-2016.

On 10th July'2016, the Portuguese beat the host France 1-0 in the final game played at the Stade de France in Paris, thanks to Eder's extra-time goal, as it ended **draw** after 90 min of play. In this Championship the greatest surprise is due to the fact that during the seven games of Portuguese in the tournament they managed to win just one match inside 90 min!. The progress of Portuguese from its first game to its last game (except one game) recorded only **draw** cases and **draw** cases of 90 min play every time!

Portugal progressed from a weak looking Group-F in third place after **drawing** their games with each of Hungary, Iceland and Austria.

In the 'Round of 16', due to a **draw** of Portugal against Croatia the game went to extra-time play after ending **draw** 0-0 before Ricardo Quaresma scored a winner goal in the 117th minute (at 27th minute of extra play).

In the next round of Portugal against Poland, penalties were needed after the game ended 1-1 **draw**.

The semi-final against Wales was the only game Portugal managed to win inside 90 min. Ronaldo opened the scoring and then provided an assist for Nani to advance to the final with a 2-0 win.

And at the Stade de France against the host France, the match, all too predictably, wasn't settled in normal time, it was **draw** case during 90 min of play. During extra time, it ended with 1-0 goal.

Besides all the **draw** cases within 90 min by Portuguese, there are many other **draw** cases too, in each of: the Round of 16, the Quarter Final, the Semi Final, and the Final!

In every UEFA Euro or even in every World Cup or even in every football championship held in this world, this type of **draw** cases are very common in the history. For example, scoreless **draw** of Brazil versus Mexico in World Cup 2014 in First Stage Group-A at Estadio Castelao Fortaleza (Brazil) on 17th June'2014, the Germany versus Ghana 2-2 **draw** game in the World Cup 2014 in First Stage Group-G played at Estadio Castelao Fortaleza (Brazil) on 21st June'2014, etc. to list a few out of many.

The author here is quite sure that on 10th July'2016 in the final match of 'UEFA Euro 2016' held at the Stade de France in Paris played by 'Portuguese versus France', if all the real time continuous data/information be input to the database in the UEFA server then the "CFE-software" can precisely compute the 'truly better' team executing its fuzzy algorithms. And surely CFE thus can give more justice to the football world as compared to the existing obsolete and weak football-rules of FIFA/UEFA and IFAB [26–29], as observed by the world on 10th July'2016 at Paris with deep sorrow mind.

Example of few more top class important games which could have been given 'better decisions' by CFE method (the information are extracted from the open-source websites [26–29]) are collected and presented below.

18.2 History of 'Penalty Shootouts' (and 'Extra Play'): Some of Them are Interesting but Some of Them Yielded Unfortunate and Shocking Final Decisions

A note on 'penalty shoot-outs' is earlier presented in Sect. 2 in this article. However, in this section some unfortunate cases of 'penalty shootouts' are presented, which can be given much better decision without using Penalty Shoot-out if CFE method were used.

The finals of many FIFA competitions, including World Cups, have gone to 'penalty shootouts'. For example:

- The 1991 FIFA World Youth Championship between Portugal and Brazil in Lisbon was decided on a penalty shoot-out which the Portuguese won.
- In the 1994 FIFA World Cup Final at the Rose Bowl in Pasadena, California, Brazil and Italy ended extra time scoreless draw. Brazil went on to win the Penalty shoot-out 3–2.
- The 1999 FIFA Women's World Cup Final between the United States and China, also at the Rose Bowl, was scoreless after extra time. The United States team won the Penalty shoot-out 5–4.
- The 2006 FIFA World Cup Final also went to a penalty shoot-out (after a 1–1 draw followed by a scoreless 30-mins. extra time) and was won by Italy 5–3 against France in Berlin's Olympic Stadium.
- The 2011 FIFA Women's World Cup Final, held at Commerzbank Arena in Frankfurt, went to a penalty shoot-out (after a 1–1 draw at full-time and a 2–2

draw after extra time) between the USA and Japan. Japan won the game after scoring 3 penalties to 1 by the USA.

- The 2013 FIFA U-20 World Cup final in Istanbul went to a penalty shoot-out after a 0-0 draw after extra time. France won the game after scoring 4 penalties to 1 by Uruguay.

Goalkeepers have been known to win shoot-outs by their kicking too. For example, in a UEFA Euro 2004 quarter-final match, Portugal goalkeeper Ricardo saved a kick (without gloves) from England's Darius Vassell, and then scored the winning shot. Another example is Vélez Sársfield's José Luis Chilavert in the Copa Libertadores 1994 finals (it should be noted that Chilavert had a reputation as a dead-ball specialist and scored 41 goals during his club career).

Antonín Panenka (Czechoslovakia) decided the penalty shoot-out at the final of the 1976 European Football Championship against West Germany with a famous chip to the middle of the goal. The English, and, to a slightly lesser extent, the Dutch and Italian national teams are known for their poor records in penalty shoot-outs. England has lost seven (out of eight) penalty shoot-outs in major tournament finals, including losses to Germany in the semifinals of the 1990 FIFA World Cup and UEFA Euro 96 (the only two times England has reached the last four of a major competition since the 1960s). Since UEFA Euro 96 England have lost five shootouts in a row in eight major tournament finals, losing to Germany at Euro 96, Argentina at the 1998 World Cup, Portugal at Euro 2004 and the 2006 World Cup and Italy at Euro 2012. The only victory was against Spain in the Euro 96 quarter-final.

The Netherlands, meanwhile, lost four consecutive shoot-outs; against Denmark in Euro 92, France in Euro 96, Brazil in the 1998 World Cup, and Italy in Euro 2000, before finally winning one against Sweden in Euro 2004. In Euro 2000, the Netherlands had two penalty kicks and four from shootout kicks, but only managed to convert one kick against Italian keeper Francesco Toldo. Frank de Boer had both a penalty kick and shootout kick saved by Toldo, who also saved from Paul Bosveltto to give Italy a 3-1 shootout victory. Penalty kick fortunes have seemed to improve during the 2014 World Cup when the Netherlands defeated Costa Rica on penalty kicks in the Quarterfinals (only to lose again on penalties in the Semi-Finals, this time to Argentina).

The Italians have lost five shoot-outs in major championships, notably being eliminated from three consecutive World Cup finals on penalties (1990–1998). However, they have also won three shoot-outs, including the Euro 2000 semi-final, the Euro 2012 quarter-final against England and the 2006 World Cup Final against France.

On 16 November 2005, a place in the World Cup was directly determined by a penalty shoot-out for the first time. The 2006 FIFA World Cup qualifying playoff between Australia and Uruguay ended 1–1 on aggregate, with Uruguay winning the first leg 1–0 at home and Australia winning the second leg at home by the same score. A scoreless 30 min of extra time was followed by a shoot-out, which Australia won 4–2.

During the 2006 FIFA World Cup in Germany, Switzerland set an unwanted new record in the Round of 16 shoot-out against Ukraine by failing to convert any of their penalties, losing 3–0. The goalkeeper Oleksandr Shovkovsky (Ukraine) became the first goalie not to concede a single goal in the penalty shoot-out saving two of the Swiss attempts with another shot hitting the crossbar. The result meant that Switzerland became the first nation to be eliminated from the World Cup without conceding any goals (and, moreover, the only nation to participate in a World Cup finals tournament without conceding a goal).

The same competition featured a shoot-out between Germany and Argentina, the two most successful teams up to that point in terms of World Cup finals penalty shoot-outs: each team had competed in 3 shoot-outs and won all of them. Germany won this shoot-out, leaving Germany alone with a 4–0 record in World Cup finals.

18.3 A Genuine Natural Question

On 20 June 2007, a new UEFA record was established. We are not sure whether we should call it a record or something else! The semi-final of the European under-21 Championships in Heerenveen between the Netherlands and England team finished in 1–1 goals. **Then as many as thirtytwo (32) number of penalties had to be taken before the tie case was settled and decided. The Netherlands eventually won '13–12' after 32 penalties by the two teams. Was it Football game or Penalty Game? Thanks to goal-keepers of both the team**.

But, does it not reflect the serious weakness of the existing 'Penalty Shootout' method of FIFA/UEFA?

19 "CFE Penalty Shootout" (CPS): A Very Very Rare Case in CFE Method

It is already mentioned earlier that there is no 'extra play' in CFE, there is no 'penalty shoot out' in CFE. However, there is a very very low possibility of applying the CPS Play in a football game under CFE method, may be once in 1000 cases of games, while TCFE is required. Because mathematically the possibility of TCFE is not absolutely zero (0).

It is earlier mentioned in Section-14 that althrough in our present article, we will prefer **TCFE-2** only (neither TCFE-1 nor TCFE-3), in case of TCFE play in CFE.

19.1 What is CPS?

CPS is the abbreviation for '**CFE Penalty Shootout**'. The '**CFE Penalty Shootout**' (**CPS**) is different from the existing concept of "penalty Shootout" practiced by FIFA/UEFA. The existing "penalty Shootout" is just a method which seems to somehow finish the game by declaring a Winner. The CPS is not so, it is truly a game of the skill of each of the 11 players of both the teams (if there is not less than 11 players).

One of the beauties of the CPS play is that it retains the football spirit of 'team performance' of all the eleven players for each team, not performance by few individuals. The "penalty Shootout" practiced by FIFA/UEFA is not a team work as the existing rule of IFAB does not allow all the players to play. The existing 'penalty shootout' method is just a test of few (maximum five) individuals which may be considered <u>inappropriate</u> in a team sport, in particular where a team size is large like 11 (eleven)!. Football is a 'team sport' but 'penalty shootout' is surely not; It is very unfortunate that in the 'penalty shootout' the football gets converted into a sport of few individuals according to the existing FIFA/UEFA rules.

Another important difference between CPS and "penalty Shootout" is that the "penalty Shootout" is a very frequent event in World Cup or Euro Cup, etc. whereas the chance of CPS play is very very low, just can not be told to be absolutely zero mathematically. The CPS play may not even be required once in 1000 cases of games!

Besides that the decision by CPS Play is not made only on the basis of number of goals scored, but also on the basis of few more parameters (positive and negative). This makes another major deviation and dominance of the CPS Play from the existing "penalty Shootout" play of FIFA/UEFA.

During CPS play, the proposed CFE method eradicates the demerit of 'penalty shootout'. Instead of existing practice of 5 kicks, the Theory of CFE introduces 10 kicks in CPS with the following strong philosophy:

The Rules for '**CFE Penalty Shootout**' (**CPS**) Play are as follows:

(i)   **CPS** Play consists of 10 penalty kicks (at most 10) by each team in the existing manner.

(ii)   with the condition that each player (except goalkeeper) of a team will kick once. None is allowed to kick twice.

(iii) Replacement of Goalkeeper is allowed by a team as many times the team wants during **CPS**.

(iv) But **CPS** does not allow replacement of any player (other than Goalkeeper) during the tenure of **CPS** play. The set of 10 players of a team (other than Goalkeeper) becomes fixed for **CPS** which is exactly the same set of 10 players of the team at the very moment while the '90 min of play' became over.

(v)   However at the time of **CPS** play, if any team is having n number of players (where n < 10) other than its goal-keeper, then the team can only make n number of penalty kicks instead of 10.

19.2 Philosophical Norms of CPS

The five important norms of **CPS** as mentioned above are a package type designed philosophically, logically and technically.

For a TCFE-2 play, the data corresponding to the 10 penalty kicks by each team will be sent to the database to execute the CFE-software. If the CS scores yet comes equal then **CPS Play** will be repeated once again and so on, untill the CFE-software computes the 'better' team uniquely. But it is really a very very rare case that the CFE after 90 min of play gets into TCFE and hence invites its 'CPS play' to compute the winner. Such a situation is not expected once in 1000 games even! Nevertheless the possibility is not absolutely zero mathematically, and hence we must have a method for the sake of soundness and completeness of our Theory of CFE.

In CPS, the 10 players will kick 10 penalty shots, none can kick twice. This makes the **CPS** different from the existing practice of "penalty Shootout". This is the most important philosophy behind the above five norms of **CPS** because of the fact that it retains the game as a game of 11 players for each team, retain much more fairness, retain much more homogeneity, retaining the football as a 'team sport' (not converting the sport into a sport of few individuals even after 90 min of play). This philosophy is <u>totally missing</u> in the existing practice of 'penalty shootout' in the FIFA/UEFA rules, which seems to somehow finish the game by few kicks.

The team to take the first kick is decided by a coin toss and the Referee himself chooses the goal at which the kicks are to be taken. All kicks are taken at one goalpost to ensure that for both the teams the kick-takers and the goalkeepers face the same pitch irregularities (if any), same wind and sun conditions, etc. as followed presently by FIFA/UEFA.

19.3 Positive and Negative Parameters in CPS (Corresponding to Each Team)

There are four parameters (for each team) in CPS Play according to the Theory of CFE. Out of four parameters, there is one **negative parameter** and three **positive parameters**. However, more number of decision contributing parameters of continuous nature, positive as well as negative, can be added for evaluation of the CPS Play in the Theory of CFE in future time, if decided by FIFA(IFAB) and UEFA.

One Negative Parameter (corresponding to each team) in CPS:

(i) BGS

Three Positive parameters (corresponding to each team) in CPS:

(ii)   G
(iii) MGS
(iv) SCB

19.4 CS Values of the Parameters in CPS

The CS values of the four parameters during CPS Play in CFE are presented below.

CS Value of the negative parameter (of a team)

(i) For every BGS, CS(BGS) = 2;

CS Value of positive parameters (of a team)

  (i)   For each Goal G scored during **CPS**, CS(G) = 10.
        (*The goals of the first 90 min of play are not to be considered in the G value of CPS, as they are now redundant*).
 (ii)   For every MGS, CS(MGS) = 1;
(iii)   For each Shot which Collides at Goal Post (Sect. 5.2) without scoring Goal during **CPS** play, the CS value is CS(SCB) = 2.
        (However, if a Shot Collides at the Goal Post with a Goal during **CPS**, then there is no CS(SCB) value for this shot as it is insignificant in this case).

19.4.1   Why Both of the BGS and MGS do not Have the Same CS Value Affixed in CPS, Although They Have the Same CS Value Affixed for the 90 min of CFE Play?

In reality, every football fan (if not supporter of any team) will expect to enjoy either a goal(G) corresponding to every penalty shot or an excellent save by the goalkeeper. A penalty shot is a game between two players only. The attacker is having an amount of freedom viz. (i) thinking time freedom (ii) freedom of using space to run for the kick (iii) freedom of making a strategically planned decision, etc. to show his skill for a success, where full span of the goal post is open to him except the goalkeeper whose role is to defend only (not to attack).

Thus, in worst case the penalty shot is **expected** to be a MGS, being a slightly missed case for goal. But, if a penalty shot becomes a BGS it can not be acceptable by heart of the fans, can not be acceptable as an added value to the spirit and interest of the game, and most important fact that it can not be acceptable to provide any good 'element of football' to the football fans (except the happiness of the players/supporters of the opposite team) at CPS play. Consequently, this negative parameter BGS in CPS deserves <u>more amount of grade</u> compared to the grade affixed for it in the CFE of 90 min of play (see Sect. 11.1). For this reason both of the BGS and MGS do not have the same CS value affixed in CPS, although they have the same CS value affixed for the 90 min of CFE play.

19.5   'CS Score' of a Team After CPS Play

At the end of CPS, the final "Winner" is to be computed on the basis of CS score of each team. CS score of both the teams are to be computed by the CFE-software. To compute the CS Score of a Team (say, team X), the CFE-software computes the following values first of all:

(i) The CS values of the negative parameter of CPS Play for the team X which is denoted by **NCS(X)**.
(ii) Total amount of CS values of the two positive parameters of CPS Play for the team X which is denoted by **PCS(X)**.

'CS Score' of the team X is then computed using the formula:

$$CS(X) = 1000 + PCS(X) - NC(X).$$

Thus the 'CS Score' out of CPS Play for both the teams X and Y can be easily computed at the server.

Note: *In the above formula for CS(X), there is no significance of the amount 1000. It is added for no other purpose but just to ensure that the value of CS(X) does not become a negative real number under any circumstances. It is obvious that even if we do not added 1000 here, the final result of CFE-software will not be changed or disturbed. If for a team X it happens at the end of CPS Play that CS(X) < 1000, then it obviously signifies that the team X is a poor team in terms of football skills and merits.*

Finally the CFE-software computes the **Winner** of this football game as explained in the next subsection.

19.6 Who is the "Winner" by CFE Method, if CPS Play be Invited?

In such a case, the CS score of both the team X and Y corresponding to the 90 min of play need not be considered now, as these two data are redundant. The decision will be taken by the CFE method on the basis of CS score of X and Y corresponding to the CPS play time only. The CFE-software will be executed in the FIFA-Server to compute the CS score of both the teams X and Y corresponding to the CPS play time only.

If CS(X) > CS(Y) then the team X is the **Winner**. Otherwise, if CS(X) < CS(Y) then the team Y is the **Winner**.

Even then a very rare case is if CS(X) = CS(Y). In that case the **CPS Play** will be repeated once more and corresponding data will be sent to the database to run the CFE-software. If the result yet comes 'draw' then **CPS Play** will be repeated once again and so on, untill the CFE-software computes the **Winner**. But these are to be thought because of the reason that mathematically the possibility is not absolutely zero(0); although in real life situations such type of cases are not expected to arise, even not once in 1000 games!.

19.7   A Hypothetical Example of a CPS Play in a Football Game of FIFA

We explain here with <u>hypothetical</u> data an example of a CPS Play in a Football Game played between two teams X and Y in a FIFA match (imaginary match). The CS scores of both the teams for the 90 min of play are equal. Consequently it is a TCFE case. Suppose that the rule is to play TCFE-2. Also suppose that the CPS play thereafter has ended with 5-6 goals after 10 penalty shots by each team. The existing method of FIFA/UEFA chooses the 'Winner' team <u>directly</u> on the basis of one and only one parameter which is the '5-6 goal score' at the end of $10 + 10 = 20$ penalty shots, ignoring the other important and significant parameters. But the CFE method is not so, it is based upon micro evaluation of all the $10 + 10 = 20$ penalty shots.

According to the Theory of CFE, the data from the field go to the database in FIFA server as the input components for the CFE-software. Suppose that the following is the statistics for the two teams X and Y as recorded in the database of the FIFA-server corresponding to the 4 parameters of CPS play, shown in Table 3.

The CS Values of 4 parameters for each team X and Y corresponding to this CPS play
The CFE-software considers the CS values of all the 4 parameters for both the teams X and Y. The execution of the CFE-software computes the CS values of all the 4 parameters for both the teams X and Y at the end of $10 + 10 = 20$ penalty shots, which are shown in the Table 4.

The CFE-software then computes the **NCS** and **PCS** values of each team as below:

NCS(X) = 0   and   **PCS(X)** = 59.
NCS(Y) = 8   and   **PCS(Y)** = 60.

The CFE-software then computes the '**CS Score**' of each team as below:

CS(X) = 1000 + PCS(X) – NCS(X) = **1059**   and
CS(Y) = 1000 + PCS(Y) – NCS(Y) = **1052**.

Final output of the CFE-software
Since CS(X) > CS(Y) by CPS play, the CFE-software outputs that the team X is the 'better' team and hence declared to be the **Winner**, the team Y is the looser.

Table 3 The statistics of 4 parameters for team X and team Y of CPS play

Serial No.	Parameters	Frequency (X)	Frequency (Y)
1	BGS	0	4
2	G	5	6
3	SCB	4	0
4	MGS	1	0

Table 4 The statistics of 4 parameters for team X and team Y of CPS play

Serial No.	Parameters	Frequency (X)	CS value (for X)	Frequency (Y)	CS value (for Y)
1	BGS	0	0	4	8
2	G	5	50	6	60
3	SCB	4	8	0	0
4	MGS	1	1	0	0

Declaration: The team X is the '**Winner**' in today's FIFA football match.

Note:

It is to be noted that at the end of the CPS play the goal results in this example is 5-6 by 10 + 10 = 20 penalty shots, i.e. the team X scored 5 goals out of 10 penalty shots whereas the team Y scored 6 goals out of 10 penalty shots. By FIFA rule the 'Winner' in this game is Y, whereas by our CPS rules of CFE the 'Winner' in this game is not Y but X. It is because of the reason that FIFA rules say Y is the better team in today's match whereas the CPS rules say X is the better team. Any football fan or football analyst will surely agree with the decision computed by the CFE-software of the CPS play.

20 Electronic Display Board: Few Sample Snapshots

Electronic Display Boards inside the stadium are used to display the latest status of performance of both the teams for information to the football fans watching the game inside the stadium and to the world fans watching the game in TV or via other media outside the stadium. There is a pair of Boards fixed as shown in Fig. 12.

They are called by Board-1 and Board-2. The Board-1 displays the real time data/information during 90 min of play. The Board-2 is to display the real time

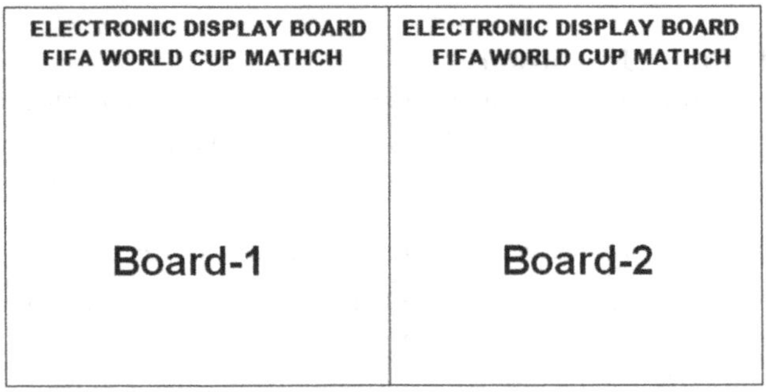

ELECTRONIC DISPLAY BOARD FIFA WORLD CUP MATHCH	ELECTRONIC DISPLAY BOARD FIFA WORLD CUP MATHCH
Board-1	**Board-2**

Fig. 12 Electronic Display Boards (two Boards)

Fig. 13 Electronic Display
Boards (Board-1)

ELECTRONIC DISPLAY BOARD
FIFA World Cup Match

Clock	Date: 12ᵗʰ January 2017 Time: 3.54 PM Play amount so far (in time) : 24 minutes Teams: Club-X vs Club-Y

Parameters	Frequency (X) : Number of Occurrences for Team X	Frequency (Y) : Number of Occurrences for Team Y
F1	3	1
F2	2	1
F3	0	0
O	3	4
H	3	2
BGS	1	0
R	3	1
G	4	3
BPC	40%	60%
BPH	26%	50%
SCB	0	0
MGS	7	6
CK	3 (with one BCK)	5
T	6	9
2G	0	0
3G	0	0
nG (n>3)	0	0

Latest CS Score (X) =	Latest CS Score (Y) =

data/information during CPS play only (if CPS play required by CFE for a match),
not of the 90 min play. However, the Board-2 will also be used for displaying
miscellaneous information, since the start of the game, whatever be decided by the
organizer.

There is a standard format configured for display of information via Board-1. In
the Electronic Display Boards, the abbreviated names and the corresponding CS
values of positive parameters will be displayed in BLUE color. But the abbreviated
names and the corresponding CS values of negative parameters will be displayed in
RED color.

The following is a sample snapshot of display while using the Board-1. This
sample of formatted display via Board-1 during 90 min of play is presented below
in Fig. 13 with hypothetical data/information.

A sample of formatted display during the 90 min of play with all details is
presented in Fig. 14 with hypothetical data/information.

In case of CPS play, a sample snapshot of of formatted display of real time data
after every kick is presented in Fig. 15 with hypothetical data/information.

The final CS scores of both the teams with the declaration of 'Winner' is to be
displayed on Board-1 at the end of 90 min if there is no CPS play. However, The
final CS scores of both the teams with the declaration of 'Winner' is to be displayed
on Board-2 if there is CPS play due to TCFE (TCFE-2). See Fig. 16 which
announces the 'Winner'.

Fig. 14 Real Time CS scores being displayed in Board-1

ELECTRONIC DISPLAY BOARD
FIFA World Cup Match
Date : 12ᵗʰ January'2017
Time : 3.54 PM
Play amount so far (in time) : 24 minutes
Teams : Club-X vs Club-Y

Parameters	Frequency (X)	CS Value (for X)	Frequency (Y)	CS Value (for Y)
F1	3	12	1	2
F2	2	12	1	4
F3	0	0	0	0
O	3	3	4	4
H	3	3	1	1
BGS	1	1	0	0
R	3	2	1	1
G	4	40	3	30
BPC	40%	2	60%	3
BPH	26%	2.6	50%	5
SCB	0	0	0	0
MGS	7	7	6	6
CK	3	5	5	10
T	6	6	9	9
2G	0	0	0	0
3G	0	0	0	0
nG $(n>3)$	0	0	0	0

Latest CS Score (X) =   Latest CS Score (Y) =

Fig. 15 Real Time CS scores of CPS play being displayed in Board-2 after every kick

ELECTRONIC DISPLAY BOARD
FIFA World Cup Match

CPS Play
(CS Score after every kick)

Date : 12ᵗʰ January'2017
Time : 5.17 PM
Teams : Club-X vs Club-Y

Parameters	Frequency (X) : Number of Occurrences for Team X	Frequency (Y) : Number of Occurrences for Team Y
BGS	1	0
G	6	3
MGS	3	5
SCB	0	2

Latest CS Score (X) =   Latest CS Score (Y) =

21   FIFA Ranking: By a New Approach

If the CFE method is incorporated in games, the future FIFA-Ranking of countries (FIFA Members) will be much more accurate. Because this method incorporates all the micro data with effect from the first match to the last match of the tournament. For example, if a team has played 10 matches, then its rank will be computed using the continuous real time data of its $90 \times 10 = 900$ min of play.

Fig. 16 Final CS scores of both the teams being displayed

```
ELECTRONIC  DISPLAY  BOARD
     FIFA World Cup Match

    Date :    12ᵗʰ January'2017
    Time :    5.17 PM
    Teams :   Club-X  vs  Club-Y

Final CS Score (X) =
Final CS Score (Y) =


        WINNER  is  Y

   Congratulation both X and Y
        for good CS Score
   Congratulation Y the WINNER
```

In a tournament, one team (say club-X) may have won 18 matches in total, which is the highest in this tournament by any team. In our proposed CFE philosophy, it does not necessarily mean that team-X is the best team among all.

Even it may happen that one team (say club-Y) may have scored 26 goals in total in this tournament, which is the highest in this tournament by any team. In our proposed CFE philosophy, it does not necessarily mean that team-Y is the best team among all.

Even in our proposed CFE philosophy, the champion team too may not sometimes happen to be of 1st Rank in the ranking.

The logical strength of this CFE philosophy can be realized by a simple example presented in the box below.

See the simple hypothetical example below (not a sports example) to understand the analogous philosophy in which a great significance and great importance of the "Second" is justified:

Example

A great significance and great importance of the "Second" is shown below with the help of a simple hypothetical example:

The four students X, Y, Z, T in a Class use to happen to become toppers (rank among first four) every year in a School. Their percentage of grand total marks obtained in consecutive five years from 2009 to 2013 are shown in the table below. The school confers a prestigious award called by "**Oscar Student of the Year**" award every year to the topper along with a certificate (Table 5).

Table 5 Results of four students X, Y, Z and T

	2009 (%)	2010 (%)	2011 (%)	2012 (%)	2013 (%)
Student X	82	75	**98**	68	71
Student Y	**98**	**96**	**97**	**98**	**94**
Student Z	72	**98**	68	**99**	72
Student T	**99**	75	82	71	**98**

From the percentage of marks (grand total), see that in the year 2009 the "Oscar Student of the Year" goes to T; in the year 2010 the "Oscar Student of the Year" goes to Z; in the year 2011 the "Oscar Student of the Year" goes to X; in the year 2012 the "Oscar Student of the Year" goes to Z; and in the year 2013 the "Oscar Student of the Year" goes to T.

One interesting point is to be noted that Mr. Y stood second in every year, probably most talented student among all of these four X, Y, Z and T, but never been awarded "Oscar Student of the Year". See that his score is very high in the school, but he is never a topper (winner). Is it a fair decision or fair selection of the best candidate adjudged just by the highest percentage of marks if considered for five consecutive years instead of one year? To remove this type of anomaly, the school has introduced "**Lifetime Achievement Oscar Student Award**" in the year 2013, which goes to Y!, not to the toppers X, Y or T.

In a played game, the CS-score of a team is a function of several parameters, some of them are positive parameters and some of them are negative parameters. These parameters have the capability to scan a team every minute during the 90 min play for evaluation by way of computation executing a software at the FIFA-server. In the philosophy of CFE, the best team must dominate in the positive parameters in the tournament having least (or almost the least) records in negative parameters. Consequently, the team having highest cumulative CS-score earned out of all the matches played by it is to be qualified to become the best or topper. However, although CPS play is theoretically a part of CFE play under certain situation, the CS-scores of CPS (if CPS required to be played in a match) is **not** counted while ranking is done at the end of all the tournament matches; only the CS-scores of its corresponding 90 min CFE play are considered and counted, although the two CS-score are equal in such a CPS case. Thus, in CFE philosophy, FIFA-ranking is to be done on the basis of '**cumulative CS-score**' earned by each team (excluding the amount earned during CPS play). This CFE method if adopted for FIFA-ranking will produce much more accurate, much more appropriate, much more scientific, much more logical, results than the existing method of FIFA-ranking.

22 Introducing 'Robot Referee' for Football Matches

The subjects like Mathematics, Statistics, Soft-Computing, Computer Engineering, Electronics Engineering, Communication Engineering, Mechanical Engineering, etc. have now reached at very advanced stages compared to those of 60–70 years back. The world can employ the rich power of modern and advanced technologies in the field of sports, in particular in Football. For excellent evaluation of football matches by one more step, we should make application of more amount of advanced technology in the **'Theory of CFE' introduced above for football matches**, in particular for input of continuous real time data directly from the field by sophisticated high resolution cameras, sophisticated fast running intelligent robots inside and outside the field as Referees (including linesmen), and having 'both ways' very fast silent wireless communication/broadcast/conference among the robots, and having 'both ways' very fast silent wireless communication/ conference of all the robots with the FIFA-server. The application of advanced technology is to be carried out considering many important facts and ideas which are mandatory for next generation football, few of which are mentioned as below:

No. 1   to identify precisely whether the goal is scored using 'God's hand',

No. 2   to identify intentional light fouls (neither F1 nor F2 nor F3) which might have produced a major undue benefit to the own team or which might have produced a major damage to the opposition team, and many other small type of odd issues, which can not/never be caught by the Referee because of the genuine reason that he is a human being whom God has given a limitation of speed and accuracy in his eyes, in his cognitive neural network, in his processing power, in his running speed, etc.

No. 3   to have visual inputs always from a near proximity of the ball during play time (video inputs of small duration). During play time, it is a frequent event that the ball suddenly from its existing locality goes/flies a long distance inside the field which is much away from the present location/ proximity of the human Referee (Referees). God has given a limitation of running-speed to the Referee (referees), to all human beings, but the ball happens many times to fly much faster compared to the running-capability of the Referee, even though the Referee is excellent in his duty, excellent in his body fitness and can run fast by his best capability. By the time the Referee runs towards the ball to reach at its proximity, the ball again sometimes goes/flies a long distance from the real time location of the Referee. But the game everywhere is being played in full-swing without waiting for the Referee to arrive at its proximity, although the two linesmen are also working for evaluation and judgment in parallel. In a football game, there are two linesmen who assist the Referee in controlling the game. The linesmen's duty is to signal to the Referee when the ball is out; to indicate a corner kick, a goal kick or to designate which team is entitled to the throw-in. The linesmen may also signal offsides, fouls or misconduct if a goal has been scored or when substitution is desired. This limitation of human being should not be ignored today in football matches because it is a subject of evaluation by one person for 22 persons along with the rules/ norms. Today we have a lot of advanced technology and scientific methods to chase behind any flying/running object in any system, to match any requisite compatibility, to take visual inputs very correctly and effectively from distance and to process them, to incorporate very fast communication system between two systems or among multiple systems, to implement and gain the benefit of parallelism, etc.

No. 4 **'Robot Referee' for Football Matches: a Proposed Model**

To understand about a tentative solution to this important and real time problem we propose a new model: an **intelligent robot**.

It is a physical machine called as '**Robot Referee**' strongly equipped with combination of lasers, high powered cameras, advanced ICT, advanced AI elements, and related hardware/softwares, etc. for refereeing the football games, even for linesmen. A 'Robot Referee' is a Referee :-

(i)   who is having the capability to look all around 360° both in horizontal and vertical directions by its multiple eyes.

(ii) who is having much more eye power and reflection than human beings in each of his several eyes.

(iii) who is having the capability to walk in the air (besides its normal work on the ground), to run/fly in the air at super-speed compared to the running-speed of human beings on earth, to remain standing in the air, depending upon the movement of the ball during play. It can reach within the near proximity of the ball most immediately inside the field.

(iv) who is intelligent with fuzzy knowledge too, who can easily work with its inbuilt 'fuzzy pocket machine' to give fuzzy input data.

(v)   and needless to mention that it will have an in-built fast computer with good memory and fast processing capability.

(vi) Also suppose that this intelligent robot can walk and can fly very fast compared to the capability of human beings, and consequently this robot can always remain in the near proximity of the ball even during the fly of the ball, althrough during the play time!

(vii) can also guarantee that there will be no physical collision of the intelligent robot (Referee) either with any of the players or with the live ball or with the Goal Post (Sect. 5.2), Goal-Area Post.

(viii) and can reach anytime very fast at the near proximity of the ball wherever it lands on the playing field after every shot, wherever it moves on the playing field, without any physical collision either with any of the players or with the live ball. etc.

Surely such type of '**Robot Referee**' (**intelligent robot**) can produce excellent inputs to the CFE-software while CFE method is applied for continuous fuzzy evaluation of the football games of FIFA/UEFA, replacing the existing obsolete method. Construction of such intelligent physical hardware robots in today's technology is not a difficult task by the mechatronic engineers in industries.

No. 5 to deploy such type of sophisticated fast running intelligent robots inside and outside the field as Referees (including line Referees), and

(a) having 'both ways' very fast wireless silent communication and conference among the robots, and

(b) having 'both ways' very fast wireless silent communication of all the robots with the FIFA-server, and

(c) having very fast processing power of video images, having the facility of video conferencing with peer robots on duty, etc.

No. 6 There is another important advantage of **Robot Referee** in football matches. The case if the **Robot Referee** during the play gets injured and replaced is not an issue at all. It is a rare case but needs to be given a thought for the sake of completeness of our soft-computing theory on football games.

Consider the case of a human Referee injured and replaced during the play. It is fact that in Fuzzy Set Theory the membership values for a fuzzy set A proposed independently by two (or more) human decision makers will be different in general. Consequently, in case of two or more human Referee the question arises: 'How to reconcile the two mode of intelligence and decisions'?

But the above discrepancy will not arise for two or more Robot Referees. One great advantage of using Robot Referee (intelligent referee) is that replacement of Referee does not effect the future intelligent decisions arising due to change in Referee. Any two Robot Referees will be exactly equally intelligent, exactly equally talented, exactly equally knowledgeable, unlike the case of two human beings (two human Referees).

In the World Cup Football matches and in all the important matches in the world, the FIFA (IFAB), UEFA and EFL can well employ the proposed machine "Robot Referee" to replace the human referee for continuous evaluation with very high precisions. In this century, the subjects viz. Mathematics, Computer Science, Software Engineering, Electronics & Communication Engineering, Soft Computing, all other Engineering branches in particular the ICT (Information & Communication Technology) etc. have reached at so high level that the cognition system of the machine "Robot Referee" can be well equipped with enormous amount of artificial intelligence and knowledge on the football norms and theories by means of minimum amount of hardware and maximum amount of required software. One of the major advantages of employing the highly intelligent "Robot Referee" in football matches is that its amount of intelligence and knowledge, its thinking capability and speed, its speed of physical movement on the field, its amount of precision in accuracy, etc. can be enhanced as much as required by FIFA (IFAB), UEFA and EFL time to time, say once every ten years, with the development of new technologies.

No. 7 By today's advanced technology, it is even possible to award graded value for CS(MGS) instead of a pre-fixed value, on the basis of merit of the shot observed at the end point while entering MGS zone. And this will be more scientifical, more logical, much fair for the interest of football. Consequently, we need to explore such type of graded CS value corresponding to each MGS in the closed interval [c1, c2] of positive real numbers, by an appropriate choice of this interval. If a shot be a MGS, then the grade for its CS value will be more if the ball is closer to Goal Post than to the Goal-Area Post. However, the grade for its CS value will be less if the ball is closer to Goal-Area Post than to the Goal Post. Obviously the CS value will be the minimum grade c1 if the ball touches(collides) the Goal-Area post without touching(colliding) the Goal Post, and will be the maximum grade c2 if the ball become a MGS after touching(colliding) the Goal post.

For example, see Fig. 17 where two distinct MGS flying shots M1 and M2 are shown. According to our hypothesis the MGS M1 will get lower graded CS value than that of the MGS M2, but surely all CS values must be in the closed interval [c1, c2].

Similarly, see Fig. 18 where also two distinct MGS low height (or ground level) shots M1 and M2 are shown. Here too the MGS M1 will get lower graded CS value than that of the MGS M2.

No. 8 However, for all BGS in a game we prefer a fixed CS value instead of graded CS values, whatever be the amount of demerits in this shot. It may be recollected from the previous sections that for CPS play the grade of BGS is different from that of BGS of 90 min play.

No. 9 and many other significant facts (which are presently being ignored in FIFA/UEFA matches), which can be considered to improve our proposed CFE method by further amount, both scientifically and technically. The sole objective is to incorporate more amount of **Transparency, preciseness, fairness, satisfaction** and **justice** to the game 'football' considering it as an important research subject for the scientists of Soft-computing, Computer Engineering, Software Engineering, Electronics & Communication Engineering, ICT, etc.

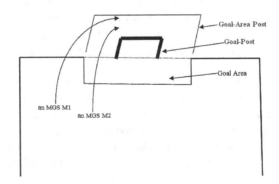

Fig. 17 The MGS M1 will get lower graded CS value than that of the MGS M2

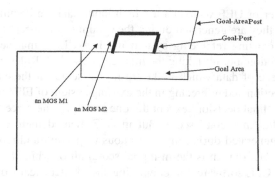

Fig. 18 The MGS M1 will get lower graded CS value than that of the MGS M2

23 Conclusion

In this work a new theory called by **"Theory of CFE"** is introduced for evaluation of any football match of FIFA/UEFA, as a replacement of the existing obsolete and weak method of FIFA(IFAB) and UEFA. The existing method of FIFA(IFAB) and UEFA was not weak in the last century (say 50 years before), but today it is weak. The weakness is explained and justified in length in this article. The CFE method is a very powerful method, very accurate method, and is constructed by a unique application of Fuzzy Theory of Prof. Zadeh in the research area of Sports Science (here it is Football sports). The abbreviation **CFE** stands for 'Continuous Fuzzy Evaluation'.

The proposed method is a continuous fuzzy evaluation method of the 90 min play to compute the 'better' team first of all, which is then declared to be the 'WINNER' of the game. The computation in CFE method is done at the FIFA-server (or UEFA-server) by execution of a software known as CFE-software. The 'fuzzy pocket machine' is a simple wireless machine M by which the inputs of the Referee are transmitted to the server directly from the playground at real time of play. For each such input, hardly 10–12 seconds of time will required by the Referee while at his duty on the playing field. The hardware of the 'fuzzy pocket machine' M can be manufactured by a good company who deals with the products on electronics and communication engineering. The code of CFE-software can be easily developed by a good programmer. It will be in fact a simple code, not a lengthy code, with very simple type of databases. The Referees need not be experts in fuzzy set theory; they can be easily trained within just 30 min of demonstration on: how to use the 'fuzzy pocket machine' to input fuzzy data to the database. Although it is called a fuzzy evaluation method, but there are crisp evaluation too in CFE method on which one (if unaware of fuzzy theory) should not be confused. The CFE method is called to be of "Continuous" nature because of the fact that mathematically the CS Score of a team is a function of several continuous variables (ex. BPC, BPH); and the final decision is computed by executing the CFE-software at the FIFA-server or UEFA-server for which input data are communicated to the server whenever they are generated during the <u>continuous</u> inspection and refereeing of 90 min. The existing refereeing system in FIFA/UEFA matches does also do continuous inspection, but for taking the final decision FIFA Rules consider the one and only one piece of data which is the 'm-n goal score' at the end of play. The continuous inspection and refereeing in the existing system of FIFA/UEFA does not contribute to the 'final decision' except the one and only one piece of data, the last data, which is the 'm-n goal' score. But in CFE method there could be a large amount of data generated during the continuous inspection and refereeing besides the last piece of data (which is the m-n goal score) all of which being input to the database of the CFE-software for computing the 'better' team of today's play, which is then declared to be the 'Winner' today.

The CFE-software will be executed nine(9) times during 90 min of play, once in every 10 min, by the real time data whatever so far been input to the server. Each of these nine executions provides the latest updated CS-score of each team which are then displayed at the 'Electronic Display Board' for information to the football fans watching the game inside the stadium and to the world fans watching the game in TV or via other media outside the stadium. Obviously, the final decision will be given on the basis of the last execution of the CFE-software i.e. the 9th execution at the end of the 90 min of play, if there is no switch over to TCFE-2 by any chance. At each occasion of these nine times display of CS score, the 'Electronic Display Board' will also display on Board-1 the break-up information i.e. the latest updated real time values of all the parameters (which are introduced in Sect. 7 in this article), and miscellaneous announcement/notice of the organizing committee (if any) on Board-2. Thus display of all latest information on the 'Electronic Display Board' will be of almost continuous type.

The categories of fouls, the list of criteria for estimating the gravity of any foul, the list of parameters, the grading values, the prefixed CS points of negative and positive parameters, Algo-1, 2, 3 for CS value of fouls by de-fuzzyfying, CS score of a team etc. may be revised or improved time to time and reset officially by FIFA (IFAB) experts and UEFA/EFL experts, and hence in our proposed CFE method they are <u>not absolutely fixed</u> for all time. The inherent scalability lying in it is its beauty to evaluate the reality.

A hypothetical example of a football game is presented to explain how the CFE can compute and produce much better decision than the decision by the existing practiced obsolete and weak norms of FIFA/UEFA/EFL. Another hypothetical example of a football game is presented to explain how the CPS play during TCFE-2 can be computed at the server by the CFE-software.

For the sake of a latest instance of weakness of FIFA/UEFA rules, one will surely agree that on 10th July'2016 in the final match of 'UEFA Euro 2016' held at the Stade-de-France in Paris played by 'Portuguese versus France', if all the continuously evaluated data/information be input then the "CFE-software" could have precisely computed the '<u>truly better</u>' team executing its fuzzy algorithms; and surely CFE thus could have given more justice to the football world as compared to the existing obsolete and poor

football-rules of FIFA/UEFA/EFL and IFAB [26–29]. There are a large number of similar cases happened in FIFA/UEFA matches in the past years, in the present and past century (few of them are presented here in Sect. 18). A new and much more accurate method for FIFA-ranking is introduced, to replace the existing ranking method followed by FIFA.

The proposed model of 'Robot Referee' can further transform the football game into a totally new era. The extraordinary power and huge advantages of such intelligent machine are the genuine need in the football matches for an excellent evaluation of continuous nature. It is the 'Robot Referees' which is to replace the present system of human Referees in future FIFA/UEFA games.

It is claimed that if FIFA incorporates this fuzzy method CFE in the World Cup Football matches and if UEFA incorporates this CFE method in the Euro Cup Football matches (in EFL matches, and other important football matches in the world) then there will be a huge enhancement in the **transparency** and **preciseness** of the evaluation, there will be a huge enhancement in the **fairness** about the decisions, because it can provide huge **satisfaction** to the football world (in particular to the looser team and their supporters), and the ultimate objective can be certainly achieved as there will be a huge improvement in the **justice** to the game 'football' too if considered as a subject of higher study, as a new era in football.

There are some games (ex. Australian rules football) where final decision is taken on the basis of score. But their method could not succeed and could not become popular to the fans too, because of the main reason that there was no element of soft-computing which can convert the impreciseness, the soft nature of reality of the game into final count. Besides that, it was not a method of continuous evaluation nature. Consequently, there was a problem of integration too, besides many other shortfalls. It was not a well-coded method for evaluation in terms of science, mathematics, statistics, engineering and technology. But undoubtedly the endeavor was good.

The proposed "**Theory of CFE**" is a theory of dynamic nature. Present version of it is the best as on today, but with the future discovery of more advanced amount of science and technology the theory needs to be improved, reviewed at least once every 10 years or so by FIFA(IFAB) and UEFA.

Acknowledgements The author is thankful to the 'Editor in Chief' Professor Janusz Kacprzyk for his valuable suggestions which have helped to improve the documentation of this book.

Future Research Work on the "Theory of CFE"

We will try in our future research work to extend the continuous evaluation method CFE to the other popular world games like Hockey, Cricket, Handball, gymnastic games, etc. But it is obvious that the games like Chess, Badminton, Table Tennis, Tennis, swimming, javelin throw, etc. to list a few only out of many, do not need fuzzy evaluation due to their althrough precise nature of play and hence a possible extension of CFE method to any of such type of games may not be appropriate.

However, to make such attempts in our future research works, a continuous encouragement and all kind of supports are required to the sports-scientists from the corresponding sports organization of the world, for football sport which are FIFA (IFAB), UEFA, EFL and FIFA members.

References

1. Atanassov, K.T.: Intuitionistic fuzzy sets. Fuzzy Sets Syst. **20**, 87–96 (1986)
2. Atanassov, K.T.: More on intuitionistic fuzzy sets. Fuzzy Sets Syst. **33**, 37–45 (1989)
3. Atanassov, K.T.: New operations defined over the intuitionistic fuzzy sets. Fuzzy Sets Syst. **6**, 137–142 (1994)
4. Atanassov, K.T.: Operators over interval valued intuitionistic fuzzy sets. Fuzzy Sets Syst. **64**, 159–174 (1994)
5. Atanassov, K.T.: Intuitionistic Fuzzy Sets: Theory and Applications. Springer, Heidelberg (1999)
6. Atanassov, K.T.: On Intuitionistic Fuzzy Sets Theory. Springer, Berlin (2012)
7. Atanassov, K.T., Gargov, G.: Interval-valued intuitionistic fuzzy sets. Fuzzy Sets Syst. **31**, 343–349 (1989)
8. Atanassov, K., Pasi, G., Yager, R.R.: Intuitionistic fuzzy interpretations of multi-criteria multi-person and multi-measurement tool decision making. Int. J. Syst. Sci. **36**, 859–868 (2005)
9. Biswas, R.: Is 'Fuzzy Theory' An Appropriate Tool For Large Size Problems ?. SpringerBriefs in Computational Intelligence. Springer, Heidelberg (2015)
10. Biswas, R.: Is 'Fuzzy Theory' an appropriate tool for large size decision problems ?. Chapter-8 in Imprecision and Uncertainty in Information Representation and Processing, in the series of STUDFUZZ. Springer, Heidelberg. (2015)
11. Biswas, R.: Introducing soft statistical measures. J. Fuzzy Math. **22**(4), 819–851 (2014)
12. Biswas, R.: CESFM: A Proposal to FIFA for a new 'Continuous Evaluation Fuzzy Method' of deciding the WINNER of a Football Match that would have otherwise been drawn or tied after 90 minutes of play. Am. J. Sports Sci. Med. **3**(1), 1–8 (2015)
13. Bouchon-Meunier, B., Yager, R.R., Zadeh, L.A.: Fuzzy Logic and Soft Computing. World Scientific, Singapore (1995)
14. Dubois, D., Prade, H.: Fuzzy Sets and Systems: Theory and Applications. Academic Press, New York (1990)
15. Dubois, D., Prade, H.: Twofold fuzzy sets and rough sets: some issues in knowledge representation. Fuzzy Sets Syst. **23**, 3–18 (1987)
16. Gau, W.L., Buehrer, D.J.: Vague sets. IEEE Trans Syst. Man Cybern. **23**(2), 610–614 (1993)
17. Gorzalzany, M.B.: A method of inference in approximate reasoning based on interval-valued fuzzy sets. Fuzzy Sets Syst. **21**, 1–17 (1987)
18. Kaufmann, A.: Introduction to the Theory of Fuzzy Subsets. Academic Press, New York (1975)
19. Klir, G.K., Yuan, B.: Fuzzy Sets and Fuzzy Logic, Theory and Applications. Prentice Hall, New Jersey (1995)
20. Mizumoto, M., Tanaka, K.: Some Properties of Fuzzy Set of Type 2. Inform. Control. **31**, 321–340 (1976)
21. Mololodtsov, D.: Soft set theory-first results. Comput. Math. Appl. **37**(4/5), 19–31 (1999)

© The Author(s) 2018

R. Biswas, *Continuous Fuzzy Evaluation Methods: A Novel Tool for the Analysis and Decision Making in Football (or Soccer) Matches*, SpringerBriefs in Computational Intelligence, https://doi.org/10.1007/978-3-319-70751-8

22. Novak, V.: Fuzzy Sets and Their Applications, Adam Hilger (1986)
23. Pawlak, Z.: Rough sets. Int. J Inf. Compt. Sci. **11**, 341–356 (1982)
24. Zadeh, L.A.: Fuzzy Sets. Inform. Control. **8**, 338–353 (1965)
25. Zimmermann, H.J.: Fuzzy Set Theory and Its Applications. Kluwer Academic Publishers, Boston/Dordrecht/London (1991)
26. http://www.fifa.com
27. http://www.theifab.com
28. http://en.wikipedia.org/wiki/Football_association
29. http://en.wikipedia.org/wiki/Penalty_shoot-out_(association_football)

Printed in the United States
By Bookmasters